FRENCH BOTANY IN THE ENLIGHTENMENT

The Ill-fated Voyages of *La Pérouse* and His Rescuers

AF772210

by

ROGER L. WILLIAMS
University of Wyoming, Laramie, Wyoming

Springer-Science+Business Media, B.V.

A C.I.P. Catalogue record for this book is available from the Library of Congress.

ISBN 978-90-481-6213-0 ISBN 978-94-017-0187-7 (eBook)
DOI 10.1007/978-94-017-0187-7

Printed on acid-free paper

All Rights Reserved
© 2003 Springer Science+Business Media Dordrecht
Originally published by Kluwer Academic Publishers in 2003.
Softcover reprint of the hardcover 1st edition 2003
No part of the material protected by this copyright notice may be reproduced or utilized in any form or by any means, electronic or mechanical, including photocopying, recording or by any information storage and retrieval system, without written permission from the copyright owner

FRENCH BOTANY IN THE ENLIGHTENMENT

ARCHIVES INTERNATIONALES D'HISTOIRE DES IDÉES

INTERNATIONAL ARCHIVES OF THE HISTORY OF IDEAS

182

FRENCH BOTANY IN THE ENLIGHTENMENT

by
ROGER L. WILLIAMS

Founders:
P. Dibon† (Paris) and R.H. Popkin (Los Angeles)
Founding Director:
R.H. Popkin (Los Angeles)
Director:
Sarah Hutton (Middlesex)
Associate-Directors: J.E. Force (Lexington); J.C. Laursen (Riverside)
Editorial Board: M.J.B. Allen (Los Angeles); J.R. Armogathe (Paris); A. Gabbey (New York);
T. Gregory (Rome); J. Henry (Edinburgh); J.D. North (Oxford); M.J. Petry (Rotterdam);
J. Popkin (Lexington); G.A.J. Rogers (Keele); Th. Verbeek (Utrecht)

Acknowledgments

The portraits of La Pérouse, D'upetit-Thouars, and Labilladière are reproduced by kind permission of the Hunt Institute for Botanical Documentation, Carnegie Mellon University, Pittsburgh, Pennsylvania.

The four maps were drawn by Ronald Hansen, Horse Creek Studio, Laramie, Wyoming, based upon maps produced by Charles-François Beautemps-Beaupré, engineer-hydrographer on board *La Recherche.*

TABLE OF CONTENTS

INTRODUCTION

The title of this book is meant to signal at the outset its confines, both its inclusions and what is beyond its scope. This volume does not stand alone, neither in splendid nor terrifying isolation, but as the third volume in a trilogy meant to be a commentary on the botanophilia that captured the literate public in 18th-century France, the very time that botany was becoming established as a science independent from medicine and herbalism.

Even though the components of that phenomenon have been examined individually by earlier scholars, a synthesis of them, founded upon a reliable command of the history of botany, has been wanting. In the two prior volumes, I discussed the discovery of plant sexuality; the foundation of a uniform nomenclature for plants and a stable, if artificial, classification by Linnaeus; the subsequent search for a more natural classification; Buffon's contribution to the popularization of natural history; the growing skepticism about the medicinal reliability of all too many herbal prescriptions that lacked experimental verification; the appeal of the senses as the key to an individual's knowledge and perception of beauty; the first publication of regional, popular floras; a burgeoning of learned societies and public libraries, primarily at the instigation of laymen committed to the cultivation of an enlightened public; the rage for the English, natural garden after the long predominance of the formal, French classical garden; the apogee of elegant flower painting; and the vigorous protests against ruinous deforestation by both botanists and agronomes. The emergence of such wide-randing natural-historical activities in France after 1750 provided enthusiastic public support for any governmental initiative likely to expand botanical knowledge. Readers seeking a more detailed account of that new cultural climate will find it in the second volume of this trilogy: ***Botanaphilia in Eighteenth-Century France: The Spirit of the Enlightenment.***

The immense curiosity about the natural world beyond Europe, and the vast number of unknown plants that had as yet escaped classification, are the subject of this third volume on botanophilia. Accounts of overseas exploration have always attracted an avid audience. It happens that several of the expeditions toward the end of the 18th century have been reexamined with commendable scholarly competence in recent years.[1] One may not only recommend such general accounts as the background for the present volume, which features only the botanists' experience on such voyages, but indicate why it has become unnecessary to reiterate anything beyond an abbreviated context. The reader interested principally in maritime, nautical, or astronomical detail will not find it here, where the emphasis is upon the botanists' response to exotic vegetation and to aboriginal peoples: their medications, their health, and their culture. Those were men who, in that day, had been first exposed to a classical education before receiving instruction in the natural sciences; and that was generally obtained during study for the doctorate in medicine. The latter combination accounts for the general assumption that the study of man was simply a part of the study of Nature. They were men who epitomized the liberalism and humanitarianism of that age with its faith in human reason and in the emancipatory powers of scientific knowledge. The latter is what they meant by liberty; and also what they meant by useful knowledge. Through them, we get insight into the mind of the Enlightenment.

That mission, as well as that of the recent more general studies noted above, is substantially at odds, in both spirit and purpose, with the fashionable, new history of science, called more precisely *sociohistory* by some of its practitioners. One of them has described traditional studies of overseas exploration for new botanical material in the 18th century as "rather precious and antiquarian in character, as having ultimately scientific rather than historical purpose, being concerned to use history in the service of systematics." Instead of which, he adds, we have "a notion which potentially unites studies of evolving natural historical practice with the social history of imperial science."[2]

That premise is made even clearer in a second contribution to the collection. Although acknowledging that Sir Joseph Banks revealed enormous enthusiasm for the discovery and identification of new species of plants, this author concentrated exclusively in his chapter on Banks as an agent of the British Empire, the implication being that 18th-century science, far from being liberal or humane in intent,

[1] John Dunmore. 1965. *French Explorers in the Pacific*, 2 vols. Oxford: Clarendon Press. Amiral Maurice D upont. 1983. *D'Entrecasteaux: rien que la mer, un peu de gloire.* Paris: Editions Maritimes et d'Outre-Mer. Catherine Gaziello. 1984. *L'Expédition de Lapérouse 1785-1788, réplique française aux voyages de Cook.* Paris: Comité des Travaux Historiques et Scientifiques. Hélène Richard. 1986. *Le Voyage d'Entrecasteaux à la recherche de Lapérouse.* Paris: Comité des Travaux Historiques et Scientifiques. Frank Horner. 1995. *Looking for La Pérouse, d'Entrecasteaux in Australia and the South Pacific 1792-1793.* Carlton, Victoria: Melbourne University Press. Brian Plomley and Josiane Piard-Bernier. 1993. *The General: the visits of the Expedition led by Bruny d'Entrecasteaux to Tasmanian waters in 1792 and 1793.* Launceston: Queen Victoria Museum.

[2] David Philip Miller. 1996. "Joseph Banks, empire, and 'centers of calculation' in late Hanoverian London." D. P. Miller and P. H. Reill, eds. *Visions of Empire; Voyages, Botany, and Representations of Nature.* Cambridge: Cambridge University Press, p. 27.

connived with wickedness and oppression. We are given no insight into Banks as a botanist with scientific views.[3] One could even conclude from the chapter that the author does not have a technical command of botany, nor does he believe that such a knowledge is necessary either to write the sociology of science or about the political uses or abuses of botany.

Such an approach may well pass muster as respectable social history. But any approach that dwells upon Banks' connivance with the promoters of imperialism and colonialism, obscuring his immense and generous services to international botany during the forty years he presided over the Royal Society, and which was anything but imperial in spirit, is quite unfair to the man and to the whole truth of the record. Such an emphasis obscures the critical association of botany with the spirit of the Enlightenment: not only the passion for new knowledge, but a knowledge to be employed for the well-being and liberation of mankind.

There is explicable reason why historians of Hanoverian Britain, rather than historians of France, should be susceptible to the imperial format. Britain was a maritime and trading nation within which a colonial policy developed quite naturally. France was continental, European in outlook, certainly interested in finding sources of precious metals, spices, timber and furs. Yet, during the later decades of the ancien régime, there was virtually no thought about finding territories overseas where surplus population could be settled. When envisioned at all, the idea of permanent settlement in distant colonies was limited to the very poor or the criminal. French navigators of that period, consequently, were not driven by a search for colonies overseas, and they had little influence on the patterns of later colonization in the 19th century. It has been too easy, and a mistake in retrospect, to equate French with British interests overseas, thus concealing the insatiable search for knowledge that imbued the French expeditions of the Enlightenment.[4]

John Dunmore publishing in 1965, could study the evidence about colonialism more dispassionately than has become customary in more recent decades. He could recognize why it had made no sense for the Old Regime, after the shocking territorial losses of 1763, to envision a resumption of imperial expansion in the face of such dominant British naval power. Only at the end of 1800, when Bonaparte regained Louisiana from Spain, did a prospectus for overseas empire reappear. The resumption of war with Britain in 1803 made the defense of Louisiana hopeless and recommended its sale to the United States. The antagonism to colonial

[3] David Mackay. 1996. "Agents of Empire: the Banksian collectors and evaluation of new lands." Ibid., p. 49.

[4] Dunmore. 1965. *French Explorers in the Pacific.* 1: 354-355.

conquest, inherent in the French Enlightenment mentality, will become manifest during our explorersí visits to Dutch and Spanish colonies.

It was well understood among botanists in the 18th century that botanical exploration on the high seas was much safer and more practical than was overland exploration on the continents. But virtually none of them could have provided the capital to outfit and man a vessel capable of sustaining an extended voyage to little-known or unknown parts of the globe. Some forty years ago, Bernard William Smith emphasized that naval vessels, whatever may have been their deficiencies, combined the value of a fortress with a traveling laboratory. Not only could the botanist travel in relative safety on a naval ship; but, when going ashore with his collecting equipment, he could enjoy the company of armed marines in the event of encounters with uncongenial aborigines, no matter that the social theory of that day predicted only high-minded, noble behavior. The naturalist could even carry a small research library on board, which facilitated the identification and preparation of specimens during the long hours between ports of call.[5]

Janet Browne was also quite aware that sea travel had accommodated naturalists more efficiently than long overland expeditions, and she quoted from Darwin's correspondence as an illustration: "I find to my great surprise that a ship is singularly comfortable for all sorts of work.–Everything is so close at hand, & being cramped, makes one so methodical, that in the end I have been a gainer...If it was not for sea-sickness the whole world would be sailors."[6]

Can there be a spoilsport so malicious as to offer an alternative to the portrayal of the Banksian plant collectors in compliant concubinage to the promoters of empire overseas? Namely, that there was no other route open to the naturalist fired with the desire to explore unknown country than to accept service on a ship of a royal navy: a free ride that paid an annual stipend, often leading to a further governmental subsidy when it came to publishing newly-discovered species and genera, along with the expensive plates that necessarily illustrated such publications.

Despite the naturalists' zeal to be appointed to such positions, it was recognized that savants and mariners, in close confinement on board ship, made poor companions. Especially on board French naval ships, the issue of whether an aristocracy of birth merited precedence over an aristocracy of brains seems never

5 Frans A. Stafleu. 1967. *Adanson, Labillardière, De Candolle.* Lehre: J. Cramer, p. 19. See B. W. Smith. 1985. *European Vision and the South Pacific, 1768-1850.* 2nd ed. New Haven: Yale University Press, a work of great distinction, originally published in 1960.

6 Janet Browne. 1995. *Charles Darwin Voyaging, A Biography.* Princeton: Princeton University Press, p. 192.

to have been conclusively resolved; but no one should be surprised that, of the two parties, the savants were the more skillfully obnoxious. Even so, the mariners greatly outnumbered them, so that the pursuit of knowledge at sea, in sum, became a perilous venture entailing heroic dedication.

The successful ideas of the French Enlightenment in the 18th century were, in fact, the culmination of ideas that had been increasingly pervasive in the intellectual culture since the 16th century: a growth of independent thinking among scientists and philosophers as a reaction against the domination of authority in prior centuries. With independent thinking came the assumption of the priority of reason over belief, accompanied by broad interest in all the natural sciences. *Nature* became the key word in the 18th century, Nature representing objectivity. Nature meant the rational: that Nature is open to reason. Whereas the supernatural meant that which is either above reason or contrary to reason. Because the deference to the natural grew out of distrust of traditional patterns of thought, that deference encouraged the search for new knowledge that could be tested as natural or rational. The essential liberalism of the Enlightenment derived from that passion to be free from traditional authority, whether theological, philosophical, or political.[7]

When one characterizes such ideas as *pervasive within the intellectual culture*, a cautionary proviso must be inserted that a substantial portion of the French population remained either illiterate or semi-literate and was not counted among the public by the measurements of that day. With that exclusion, the assertion by one especially qualified observer in 1797, Milet-Mureau, that there was general *public* interest in the progress of the physical and natural sciences, seems undeniable.

As for those who read the reports about overseas expeditions in that day, individual motives varied greatly. Some read only for diversion. Others simply took pride in the comparisons between European practices and manners with those of the aborigines, confident in the superiority of civilized man over all others. Only the learned, especially the scientists, studied such materials solely for the purpose of extending their knowledge. The learned are instinctively drawn to anything new or extraordinary, beyond which the reports allowed them to share the perils, the difficulties, and the pleasures met by the navigators along with the marvelous objects collected over vast distances. True intellectual excitement![8]

[7] Stafleu. 1967. *Adanson, Labillardière, De Candolle*, pp. 6-7.

[8] Louis-Antoine Destouff, baron de Milet-Mureau, ed. 1797. *Voyage de La Pérouse autour du monde*. 5 vols. Paris: Imprimerie de la République 1: xvii-xviii.

No hint of colonialism whatever, at least in the traditional sense of that word. Take for example the observation of a French naturalist with the Baudin expedition that had visited Australia. "Destiny," he noted, "has deprived the inhabitants of Australia of ample nourishing plants, and thus to be forever barbarian. The Europeans, as members of a common humanity, have an obligation to bring a knowledge of food plants to such people, thereby also introducing the seed of civilization."[9] Not until the recent decades of our own era would any readers find more insidious implications in such reports.

The recent critic is likely to view such a voyager's observations as nothing more than a version of class, racial, or gendered prejudices, fostering either *visions of domination* or as assertions of Western superiority to all other cultures, all of them seen as more or less savage. The voyagers' reports, in other words, no matter how scientific or sympathetic they may have been, really served Western colonialism. Even anthropologists have begun to worry that their research into non-Western cultures, no matter their scientific purposes or their sympathetic views, has the effect of serving colonialism.[10] In such an intellectual climate, the traveling-botanist of the 18th century has to be recalled as a crypto-colonialist. His apparent humanitarianism has been recast to be an egotistical desire to demean and to subvert. But as he had no inkling of such motives, he could pursue his research overseas in the expectation that he would be remembered gratefully for the hardships and perils he endured over long months and years.

The deprivations one suffered during such lengthy expeditions were more than compensated for by their incomparable intellectual benefits. Hunter Dupree's observations on the matter, made in the context of Asa Gray's last-minute decision not to be the principal botanist for the expedition to the Pacific and the South Seas authorized by Congress in May of 1836, apply to the 18th century as well as to the 19th. Gray did so for understandable reasons given the lengthy political bickering that preceded the eventual sailing of the expedition under Commodore Charles Wilkes from Hampton Roads on 18 August 1838. (It did not return to New York until 10 June 1842.) But Gray also did so in the distressing knowledge that his resignation would leave William Rich, a political botanist, an amateur at best, in charge of botanical exploration, a man hardly the equal of James D. Dana, the zoologist and geologist.

[9] Jean-Baptiste-Louis-Claude-Théodore Leschenault de La Tour. "Notice sur la végétation de la Nouvelle-Holland et de la Terre de Diémen." François Péron. 1807-1816. *Voyage de découvertes aux Terres Australes... pendant les années 1800-1804.* 2 vols. atlas. Paris: Imprimerie Impériale. 2: 359.

[10] Harry Liebersohn. 1996. "Recent Works on Travel Writing." *Journal of Modern History* 68 (September): 618.

Even in the knowledge that Asa Gray would have a distinguished career ahead of him despite that decision, Dupree correctly insisted that it had been a mistake not to participate in an expedition that did make scientific history. Exploring expeditions were the graduate schools for many of the best natural scientists of the 18th and 19th centuries: Robert Brown, Alexander von Humboldt, Charles Darwin, Alfred Russell Wallace, Aimé Bonpland, and Joseph Dalton Hooker. In both centuries, such expeditions were mainly borne in naval vessels that served a combination of a nation's commercial and scientific interests, and the science was by no means betrayed by that linkage.

In Asa Gray's case, we cannot know precisely what such an exploratory opportunity would have contributed to his interest in the geographical distribution of plants. Neither can we believe it would have been negligible. Without doubt, his resignation made him increasingly vulnerable in later years to the charge of being a *closet botanist*, meaning that everything was studied second-hand and suggesting pedantry, fairly or not. The charges would be made in particular by some specialists in the flora of the Rocky Mountains, who, like their 18th-century predecessors, had had the advantage of examining their plants in their habitats before a later reexamination in dessicated form.[11]

In France, the era of the lengthy sojourn overseas for botanical collections began not in the 18th century, but in the 17th. But the character of such botanizing in the 17th differed in several senses. For the most part, its major figures were members of religious orders, they had missionary obligations, and they were often preoccupied to find plants with medicinal virtues.

The publication of *Genera Plantarum* by Linnaeus in 1737, in which he first exposed his ingenious sexual system for plant classification, can be regarded roughly as a turning point. Thereafter, botanists traveling abroad were predominantly laymen, largely indifferent, sometimes hostile, to proselytizing. And, although far from being unconcerned about the utility of plants, they were emphatically concerned with the discovery of new genera and species and their placement in a stable classification. We are at the birth of botanophilia.

This new pattern of French botanical exploration becomes apparent in the career of Michel Adanson, who, at the age of 20, left for Senegal in 1748, having obtained a position in the Compagnie des Indes. As a clerk, he had to spend much of his time in Fort-Saint-Louis, going only occasionally to visit other company offices in that country. He was not, properly speaking, an explorer, but rather a sedentary employee intellectually isolated in a country where everything remained

[11] A. Hunter Dupree. 1988. *Asa Gray, American Botanist, Friend of Darwin*. Baltimore & London: Johns Hopkins University Press, pp. 58, 68-69.

to be identified and classified. Yet, after six years, he landed at Brest in 1754 with thousands of botanical and zoological specimens, with notes on the geography of Senegal and the customs of the local population. All of the field work had been accomplished on his own initiative and without any governmental subsidy. He had returned convinced, moreover, that no extant classification system could accommodate the immense variety of new species that had come to his attention.

His immediate task was to prepare his *Histoire naturelle du Sénégal* for publication, which appeared in 1757, a remarkable work that opened the doors of the Académie des sciences to him at the age of 30. His larger task was the development of a proposed natural classification method, which he explicated in vast detail in his *Familles des plantes*, published in two parts in 1763-1764. A highly idiosyncratic work and impractical in several respects, the book nevertheless made remarkable contributions to the development of a workable natural classification and demonstrated the results possible from extended observations in the field overseas.

Botanists today are unlikely to recognize the name Louis-Guillaume Lemonnier. He published little of interest to them, in truth, not much at all. Yet, it would be hard to overstate the importance of his career to the progress of botanical research in the later 18th century. In his own time, he was revered, much as was Malesherbes, as the selfless promoter of botanists who required major financial assistance to carry out botanical exploration overseas. His authority depended, first, upon his technical competence in botany; and second, upon his firm credit with the crown, initially with Louis XV, later with Louis XVI.

Trained as a physician (and in botany by Bernard de Jussieu), Lemonnier was serving as a physician in the royal infirmery at Saint-Germain-en-Laye when a chair was vacated at the Jardin du roi by the death of Antoine de Jussieu in 1758. Only after Bernard de Jussieu refused the appointment did Lemonnier accept the chair. Later, in 1786, the chair would be transferred to René-Louiche Desfontaines when Lemonnier took up residence at Versailles as physician to the king.

In the meantime, Lemonnier had encouraged Louis XV to develop an experimental garden at the Petit Trianon and to put Bernard de Jussieu in charge. There, under Bernard's direction, the first garden to demonstrate a natural classification was planted in 1759. Thereafter, Lemonnier's growing friendship with the king enabled Lemonnier to become a patron of botanical exploration, easily obtaining funds from the king when an appropriate candidate merited help. The benefits were, in truth, mutual, as Lemonnier cultivated a personal garden at Versailles where he worked to acclimate fine exotic trees and plants from seed received from various parts of the world, hoping to provide useful plants for French fields, gardens, and forests.

Beyond his personal general herbarium of 10,000 specimens, he obtained more than 5,000 plants from Commerson's voyages; about 1,000 plants from Labillardière's Syrian trip; around 600 plants from Desfontaine's sojourn in North Africa; and seeds especially from André Michaux's trips to Persia and North America. His collection was acquired by Benjamin Delessert in 1805 and, consequently, is preserved in the great herbarium in Geneva.[12]

Following the death of François de Lassone in 1788, Lemonnier assumed the position of first physician to Louis XVI. He had by then refused all fees from his private practice because of his official salary. On the terrible day of 10 August 1792, he was living in the Tuileries and probably owed his life to the widespread respect he had earned. It appears that one of the leaders of the assault upon the palace that day recognized him and escorted him across the Pont Royal to the safety of the Luxembourg Palace.

His life was saved, but the fall of the monarchy cost him his position and fortune. All he possessed was a personal library, which he could not bring himself to sell, and a small property in Montreuil, a town just east of Paris. He was reduced to opening a small herbal shop in Montreuil, in which he lived, selling medications and giving out the best medical advice he could, grief-stricken over the fate of former friends and the fine trees he had planted. He died there at 82 in 1799: from botanist to the king to wretched herbalist.[13]

Philibert Commerson, whose inability to publish the results of his vast botanical research overseas has led to near oblivion, was another physician who made contributions of first-rank importance to the perfection of botanical science in the later 18th century. Born to a seigneurial family that meant for him to go into law, Commerson acquired such a taste for natural history as a schoolboy that it was finally agreed that he could go to Montpellier for a degree in medicine, where he resided from 1747 to 1755. He early showed himself to be an ardent worker of the sort who pushes every interest into a passion. Even as a medical student, he began to form his own herbarium, which became a mania to make it the largest and rarest of all known herbaria. He let nothing stand in his way, raiding the botanical garden and hothouses at Montpellier to the despair of the faculty and head gardener, until he was forbidden access to those facilities.[14]

Commerson used Montpellier as his base for nearly four years after completing his medical doctorate, botanizing in the Pyrenees, the Cévennes, in Provence,

12. Antoine Lasègue. 1845. *Musée botanique de Monsieur Benjamin Delessert*. Paris: Fortin, Masson, pp. 53-60.
13. Cap (Paul-Antoine Gratacap, called). 1854. *Le Muséum d'histoire naturelle*. Paris: L. Curmer. 1: 32-34.
14. Charles-Frédéric Martins. 1854. *Le Jardin des plantes de Montpellier*. Montpellier: Chez Boehm, p. 39.

and into the Alps. In 1755, he passed into Switzerland for the purpose of meeting Albrecht von Haller, then descending upon his native Burgundy, to botanize of course, but also to establish a medical practice. By then, word of his enormous energy as a collector had reached not only Haller and Linnaeus, but Bernard de Jussieu as well; and Jussieu put out the word that Commerson belonged in Paris.

Even so, Commerson had been making an effort to live a stable, settled life. About 1758, he established a small botanical garden in Châtillon-sur-Chalaronne and worked to put his collections in order. Two years later he married, but his great happiness was cut short in 1762 when his wife gave birth to a son that cost her her life. Whether it was that incident, or his freely-expressed doubt about the efficacy of medicine as practiced in that day, that caused him to abandon the region, he moved to Paris in 1764. Given his substantial experience in the field, he soon became the obvious botanist to go on the scientific expedition by the Navy to the southern hemisphere under the command of Captain Louis-Antoine de Bougainville.

The expedition sailed in 1766. Commerson, then 39, produced the plan, or guide, for research in natural history that would serve as a model for later voyages as well. After crossing the Atlantic to South America, the expedition passed through the Strait of Magellan en route to Tahiti. All of Commerson's Tahitian collections were ultimately lost or fell into other hands. After 21 months, Bougainville reached Mauritius, the base for what would be Commerson's greatest work. The royal intendant for the colony (then called the Ile de France), Pierre Poivre, himself an experienced voyager and naturalist, had received instructions from the ministry of the marine to persuade Commerson to leave the expedition at that point so that he might explore both Mauritius and Madagascar.

He agreed to remain, if reluctantly because of fatigue after so many months at sea, which no doubt contributed to his vulnerability to various tropical diseases. Despite illness, he moved on from Mauritius to Madagascar in 1770. He found the island to be admirable, a botanist's paradise: thousands of unnamed species! That spectacle led him to wonder how any reliable classification system could be formed with so much material there and elsewhere still unknown. After four months on Madagascar, he moved on to Réunion (Ile Bourbon) early in 1771 for exploration before returning to Mauritius toward the end of the year.

At that point, Pierre Poivre was recalled as intendant, and it is unclear whether Commerson was by then too ill to accompany him, or whether the ship bearing Poivre could not accommodate the immense collection Commerson had acquired by then. Thus, he remained on Mauritius with his plants. It appears that he had a foreboding that, while his collections would be retrieved eventually by the Navy, he himself was doomed. In sending a package of plants and seed to his patron,

Lemonnier, Commerson wrote: "I have only one request to make of you. Have my priority of date conserved for those things that are truly new, which you may pass on to others. In your hands, I know the security of my deposit. But allow me to say that, in the Republic of Letters, as in the beehives, the lazy, unoccupied drones depend solely upon the active and the industrious bees."[15]

Evicted from the Intendance by Poivre's successor, a man indifferent to the sciences, Commerson, ill and grief-stricken, lost all interest in his work and succumbed on 13 March 1773 at the age of 46. The ministry of the marine, quite aware of the value of Commerson's material, lost no time in ordering its recovery. In 1774, the 32 chests containing his manuscripts and collections were delivered to the Jardin du roi in Paris. The king's decision to entrust the entire Commerson herbarium and his field notes to Antoine-Laurent de Jussieu for analysis no doubt reflected Lemonnier's advice and became of fundamental importance to Jussieu during his preparatory work on the *Genera Plantarum* (1789).

The original set of plants in Paris numbered 30,000 specimens and was recognized as the largest collection ever acquired by a single individual. It was determined in time that the collection contained about 5,000 species, of which 3,000 were entirely new to European knowledge. In later years, Jussieu would publish 37 of Commerson's new genera with proper attribution, an appropriate example being *Bougainvillea* Comm. ex Juss. in the Nyctaginaceae; as well as publish another 15 genera based upon Commerson material.[16]

Commerson's chests also contained about 1,500 drawings of plants and animals from nature by Jossigny, a young painter in his service. Beyond which Commerson had written observations on other fields within the natural sciences and, in particular, upon the peoples native to Tahiti and Madagascar, the latter being the aboriginal pygmies, presumably the Vazimba. His literary style was concise, clear, and often witty, the work of a learned man entirely at home in Latin. During his last years on the African islands, he had also managed to send packages of plants and seed to his benefactor, Lemonnier, which accounted for an additional 5,000 specimens.

In 1773, the year of Commerson's death, René-Louiche Desfontaines, a young Breton, came to Paris seeking a medical degree. He had come from an impoverished family, and his performance as a schoolboy had been anything but promising. Yet, at the Collège de Rennes, he had caught fire and advanced to the head of his class. In Paris, he followed the lectures of Félix Vicq d'Azyr,

15. Cap. 1864. *Etudes biographiques pour servir à l'histoire des sciences*, 2 vols. Paris: V. Masson. 2: 90-129.

16. Frans A. Stafleu. 1964. "Introduction." Antoine-Laurent de Jussieu. *Genera Plantarum*. Weinheim: J. Cramer, p. xi.

the eminent physician and anatomist, but could not overcome his revulsion to anatomical research. The classes at the Jardin du roi were considerably more to his taste, where his ability and zeal attracted the attention of Antoine-Laurent de Jussieu. The latter advised him to introduce himself to Lemonnier who could be very useful to aspiring botanists.

While Desfontaines did complete work for the medical doctorate, his dedication had been redirected to botany. His election to the Académie des sciences at the age of 33 (1783), after having read only several papers before that body, and before he had published anything major, surely suggests a benevolent influence on his behalf. And, indeed, that very year Lemonnier obtained subsidies for him to undertake an expedition overseas. With the approbation of the Académie, he would spend the next two years (1783-1785) exploring the western coast of North Africa, called Barbary in those days. It took him some years thereafter to prepare his material for publication, but it appeared in two volumes in 1798-1799: *Flora atlantica, Agro tunetano et algeriensi crescunt* (the Atlas, Tunisia, and Algeria).

The work provided descriptions of 1,520 species, 300 of which, plus 6 genera, were new to science, along with 261 plates. Some of the plants therein, incompletely described, were subsequently republished and dedicated to him. Beginning with his North African collections, Desfontaines gradually assembled a very large personal herbarium, drawing upon duplicates from the collections of others from various parts of the globe. His herbarium contributed greatly to the advancement of systematics, becoming a valuable element in the collections of Paris, Florence, Geneva, and Neuch,tel.

In 1786, after Desfontaines's return from North Africa, Lemonnier persuaded Buffon, the intendant of the Jardin du roi, to transfer Lemonnier's professorial chair to Desfontaines. As one of the royal physicians, Lemonnier did not need the income from a chair, which could provide Desfontaines with an income for the remainder of his career. Consequently, Desfontaines largely devoted himself henceforth to his teaching obligations, giving public lectures on botany three times a week, generally attended, in that heyday of botanophilia, by an audience of 500 to 600.[17]

Desfontaines did not suffer inordinately during the Revolution that dislocated or destroyed so many professional careers. He had become so hardworking and withdrawn that he seems hardly to have perceived the dangers outside the Jardin du roi, rebaptised the Jardin des plantes. It was believed that he had emerged from his studious retreat once to visit and console the unfortunate naturalist,

[17] Adrien Davy de Virville et al. 1954. *Histoire de la botanique en France.* Paris: Société d'Edition d'Enseignement Supérieur, pp. 80-82, 108.

Louis-François Ramond de Carbonnières, the constitutional-royalist who had been obliged to relinquish his seat in the Assembly after 10 August 1792; and once again in 1794, in conjunction with André Thouin, to appeal for the release of L'Héritier de Brutelle. Whether their intervention played any part in securing his release remains obscure.[18]

The point is, however, that having been footloose for so many months in potentially dangerous country, Desfontaines had settled into a house within the Jardin des plantes from which he rarely emerged for the remainder of his life. It would not appear that he had a fear of other people as he graciously lectured regularly to substantial audiences. What is more, he became an habitué of André Thouin's modest home on the grounds of the Jardin where a small group of naturalists gathered most every evening for informal sociability. To all appearances, he had simply ceased to be concerned for the world beyond the garden walls.

Before turning to botanical exploration in greater detail in succeeding chapters, it is germane to observe that Desfontaine's apparent eccentricity following a lengthy exploration was far from unique in the history of explorer-collectors. American botanists may recall the example of the English botanist, Thomas Nuttall, who, for the decade that he was curator of the botanic garden at Harvard, was extremely retiring and never became part of the Cambridge community.

Hunter Dupree, without citing other examples, remarked that many great travelers have been uncomfortable, *ill at ease*, in communities or settlements.[19] The question may legitimately arise as to whether they became great travelers because of that quality, or because they never again adjusted to settled life and its formalities after long periods of enforced self-reliance. The evidence furnished from the lives of additional explorers strongly suggests that the latter explanation–the long periods of enforced self-reliance–has the greater merit.

Paul Horgan became conscious of the phenomenon through his study of Josiah Gregg, a businessman who had been in delicate health from childhood, but who had regained complete health in the 1830s through participation in the Santa Fe trade. Apart from being successful in business, Gregg turned every trading trip into a scientific expedition. He made collections, made notes, and kept diaries. Once his trading days were over, he wrote: "Since that time, I have striven in vain to reconcile myself to the even tenor of civilized life...and have sought in its amusements and its society a substitute for those high excitements which have attached me so strongly to prairie life...Nor do I find my taste peculiar; for I have

18. Cap. 1854. *Le Muséum d'histoire naturelle.* 1: 49-50, 59.

19. Dupree. 1988. *Asa Gray,* pp. 104-105. See also Jeanette E. Graustein. 1967. *Thomas Nuttall.* Cambridge: Harvard University Press, pp. 176-292.

hardly known a man, who has ever become familiar with the kind of life which I have led for so many years, that has not relinquished it with regret." He added elsewhere, moreover, that his tour of the prairies had not been as dangerous to life and health as "prejudicial to his domestic habits." From which Horgan concluded, correctly, that propriety is the enemy of curiosity; that even though such explorers and collectors had no desire to appear eccentric, in society they became alien and were judged by others to be puzzling creatures.[20]

This matter took on an additional dimension within the botanical community itself. The history of botanical science surely suggests that certain aspects of research are apt to render their practitioners ill-fitted for any other kind of work. Botanical explorers in particular, drawn by the irresistable charms of exploration, have often shown little patience for the laboratory work necessary to determine their collections. That work may revert to the sedentary botanists who have never seen in Nature the material they are describing, a limitation conducive to error.[21] Or, the actual collector may undertake the uncongenial work himself, given the leisure of his likely social insulation, but take an inordinate amount of time to bring his discoveries into print as a consequence.

In French sources, one frequently sees references to Commerson as having suffered a martyr's death, succumbing prematurely on a remote island, having expended his life for the advancement of botanical knowledge. That emphasis has effectively minimized the sacrifices by botanical contemporaries, who, if subjected to perils and hardships no less than his, were fortunate enough to survive to live out their days in France. Yet, their lives were altered forever after, monastic if not formally cloistered; and they rarely married or had children. In that respect, although they were laymen, they remained within the tradition of their precursors of the 17th century who had been religious. But their faith was in human reason, for which they dedicated their lives.

[20] Paul Horgan. 1941. "Introduction." M. G. Fulton, ed., Josiah Gregg. *Diary and Letters*. Norman: University of Oklahoma Press, pp. 15-21, 36.

[21] Davy de Virville. 1954. *Histoire de la botanique en France*, p. 10.

II
Royal Instructions for Scientific Exploration

Of the three overseas expeditions that we shall examine in greater detail below, only the two major ones were sponsored by the royal régime and, therefore, guided by extensive instructions that specified the scientific information desired by both government and the premier academic societies in Paris. Charles-Pierre Claret de Fleurieu, the naval officer who was the chief architect of the instructions, drew upon three major sources of information: the recommendations that Commerson had made at the time of the Bougainville expedition; the records of prior voyages, most notably the observations by the naturalist William Anderson, who accompanied the second Cook voyage as surgeon, 1772-1775; and voluminous recommendations that Fleurieu solicited from the Parisian academies. Louis XVI, it must be added, took a pronounced personal interest in promoting such expeditions, as he had a passion for geographical knowledge. He made occasional recommendations of his own, and his personal interest meant generous financial support.

The three expeditions had in common an affiliation with the first of them, the voyage of La Pérouse, which sailed from Brest in 1785 and would be lost at sea in 1788. The second expedition, organized to search for La Pérouse and to fulfill his scientific mission, was commanded by Entrecasteaux, departing in 1791. The third voyage, led by Dupetit-Thouars, was a privately-financed venture in search of La Pérouse and botanical knowledge, sailing in 1792.[22]

The scientific relationship of the La Pérouse and the Entrecasteaux voyages becomes transparent from their official instructions. Each set of instructions differed in its assigned route for the evident reason that La Pérouse was to travel around the world; Entrecasteaux was to go to La Pérouse's last known location and then follow his presumed route. But the remaining segments of the instructions, directing various aspects of the scientific mission, were identical for both voyages.

22. As the French Revolution would intrude upon members of several of these expeditions, forcing a change in the spelling of aristocratic or seigneurial names, I have elected to use the post-revolutionary spelling unless the individual died before the Revolution.

The king's participation in the preparation of scientific expeditions had been stimulated in particular by the great interest he had taken in reading James Cook's *Voyages.* He had been moved to have a special edition on South Sea voyages prepared for the education of the dauphin. The account of Cook's third voyage inspired him to plan what came to be the La Pérouse voyage, which he hoped would impart a special character to his reign.[23]

His own character may be seen in his personal directive to La Pérouse: "At all times, le sieur de La Pérouse will deal with the different peoples he will visit in the course of his voyage with great kindness and humanity, and will devote himself with zeal and interest to every means that can improve their condition by providing their countries with the legumes, the fruits, and other useful trees from Europe; and by teaching them the way to seed and cultivate them. By making known the uses they must make of these presents, the goal is to increase, on their own soils, the necessary productions of those peoples who draw their nourishment from the earth."[24]

Even though Fleurieu had valuable sources available to him when preparing the instructions for an expedition, he was a man of considerable experience as an hydrographic engineer, a naval captain, and a naval administrator. His writing testified to his knowledge of the sciences of his day, but he knew himself to be more at home in matters of navigation, astronomy, and foreign policy than in the biological sciences. He acknowledged that he had *l'esprit mécanique*, essentially meaning *l'esprit astronomique,* and tended to appreciate the observations of his naval naturalists for their usefulness in the practice of navigation. What did the presence of certain plants, birds, or shells indicate about landing conditions? or surface currents? or geographical positions?

That priority should not cause us to overlook formative instruction he had received when still young in Lyon. Fleurieu had originally been destined for a clerical career. Along with his brother, he had been tutored by the Benedictine Dom Antoine-Joseph Pernetty, whose true passion was botany. Pernetty's influence must have been substantial. The older brother, Marc-Antoine Claret de la Tourette (later de Fleurieu), although going on into law, devoted himself to botany in Lyon, becoming a close colleague of the eminent agronome, abbé Jean-François Rozier. Given the fact that Pernetty would later leave the Benedictine order, having served as a chaplain for the Bougainville expedition, one must suspect that

23. Bernard William Smith. 1985. *European Vision and the South Pacific,* 1768-1850, 2nd ed. New Haven & London: Yale University Press, p. 137.

24. Yvonne Letouzey, ed. 1989. *Le Jardin des Plantes à la croisée des chemins avec André Thouin, 1747-1824.* Paris: Muséum National d'Histoire Naturelle, p. 198.

he could have nudged Charles-Pierre toward a career in the world, especially in the navy. In any case, Fleurieu certainly had had instruction in botany.[25]

Fleurieu's correspondence with the king by 1785 clearly indicated that overseas expeditions would serve both French commercial interest (notably the fur trade) and scientific discoveries. But Fleurieu also emphasized that the commercial interest would only be an *accessory* to a campaign of discoveries. They could be complementary objectives, but absolute priority was reserved for science. The instructions for those scientific objectives revealed that, in Fleurieu's mind, the nautical sciences, leading to a new cartography, were given clear priority over the physical, natural, and human sciences, hardly surprising given his background. Jean-Nicolas Buache de la Neuville, first geographer to the king, was given the responsibility for preparing an inventory of previous expeditions to accompany the instructions for La Pérouse, along with maps of some use, imprecise though they were and in need of perfection: a major objective of the voyage.[26]

La Pérouse's route was prescribed in Part I of his instructions, which revealed the French desire to complete the work of James Cook by exploring places Cook had been unable to cover: namely, islands in the South Atlantic, previously reported but not verified; the northwest coast of America from Monterey to Mt. St. Elias (an important source for furs); the northeast coast of Asia from Korea to Japan and Siberia; the Solomon Islands; and the king added the southern coast of Australia as a region unknown.[27]

Catherine Gaziello, in her excellent reference work on all aspects of the La Pérouse campaign, directly confronted the charge, made by some scholars, that the route suggested that the heralded scientific objectives of the voyage had merely been a screen to conceal the real goals of the expedition so as not to awaken the suspicions of the anti-colonialist party in France.[28] Yet the entirety of Part II in the instructions quite openly spoke of seeking trading opportunities for French merchants abroad but as a side effect of the expedition.[29] Louis XVI himself had made scientific discovery the primary motive for the expedition. Most convincing of all would be the conduct of La Pérouse during the voyage, which would clearly point to the primacy of intellectual inquiry.

25. Pierre M. Niaussat. 1989. "Fleurieu et les naturalistes de son temps." Ulane Bonnel, ed. *Fleurieu et la marine de son temps*. Paris: Economica, pp. 162-164.

26. Jean Bourgoin. 1989. "Fleurieu et le voyage de Lapérouse." Ibid., pp. 64-66, 82.

27. The details may be found in Louis-Antoine Destouff, baron de Milet-Mureau, ed. 1797. *Voyage de La Pérouse autour du monde*, 5 vols. Paris: Imprimerie de la République. 1: 14-29.

28. Gaziello. 1984. *L'Expédition de Lapérouse 1785-1788*, pp. 54-55, 61.

29. Milet-Mureau. 1797. *Voyage de La Pérouse*. 1: 30-42.

Gaziello had also studied material on La Pérouse's private life, not published until a century after his death, which convinced her of his well-developed humanitarian sensibilities. Before the departure of the expedition, he had seriously pondered what should be the European's proper conduct in the presence of aborigines, revealing profoundly philanthropic instincts. It had been known, moreover, that he had earlier demonstrated those instincts in his treatment of defeated British troops during his campaign in Hudson's Bay.[30] It strains credibility to suppose that such a man would have been chosen to command a clandestine operation meant to subvert the independence of aboriginal peoples, or that he would have accepted such an assignment.

On the contrary: everything regarding the selection of personnel for both the La Pérouse and Entrecasteaux voyages indicates scrupulous concern to provide professional personnel qualified both mentally and physically to fulfill those missions stated in the instructions. On the two ships under La Pérouse's command, there would be 10 scientists and artists, along with a gardener, horologer, 2 hydrographic engineers, and a Russian interpreter. If the 15 were fewer in number than the naval officers (19) on the two ships, they remained an important group on each ship whose needs required the attention of the commander-in-chief.

He did not select the scientific staff. That was left to the Académie des sciences; but he did urge that, beyond their scientific competence, the candidates' youth, enthusiasm, and compatibility should be taken into account, especially the latter quality given the probable length of the voyage under confined circumstances.[31] As anyone who has recruited personnel with rigorous care can testify, a prior assurance of compatibility will infrequently square with later performance. In the case of the gardener, choice was delegated to André Thouin, the chief horticulturalist at the Jardin du roi, as well as the task of providing his instructions. It would be the first important official expedition for which Thouin helped with arrangements, occupying him during May and June of 1785.[32]

We turn next to extracts from those documents within the instructions prepared for both La Pérouse and Entrecasteaux that described their assignments in both botanical and human sciences, as well as the conduct recommended to them in dealing with aboriginal or native peoples met en route. It must be understood that their instructions covered other intellectual fields as well, but which are beyond the province of this book.

~

30. Gaziello. 1984. *L'Expédition de La Pérouse*, p. 81.

31. Ibid. p. 138.

32. Letouzey. 1989. *Le Jardin des plantes*, p. 198. Emma C Spary. 2000. *Utopia's Garden: French Natural History From Old Regime to Revolution* (Chicago: University Press), published after the completion of this trilogy, provides an account of Thouin's vast horticultural correspondence with French botanists.

(1)

Letter from the secretary of state for the Navy, Charles-Eugène-Gabriel de La Croix, marquis de Castries, maréchal de France, to the marquis de Condorcet, permanent secretary of the Académie des sciences.

Versailles, March 1785

Following the king's decision to dedicate two frigates for a voyage that will be useful both for his service and to provide an extended means of perfecting the knowledge and the description of the terrestrial globe, the minister would like the Académie des sciences to prepare a memoir recommending in some detail which physical, astronomical, geographic, and any other observations that are appropriate and important to be made, whether at sea when navigating, or on lands or islands that might be visited. In circumnavigating the globe, the itinerary will range between the 60th parallels north and south. Consequently, the Académie may include nearly the entirety of known coasts and islands in its speculation, as well as the entire ocean surface between the great land masses that form the continents.

The king is greatly interested in the work you are asked to do, and you may be sure that he will give the closest attention to the observations or experiments you will recommend to him. His Majesty will be pleased, moreover, to see the luminaries of the Académie des sciences manifesting the same zeal and love of glory that animates the officers of the Navy, promising the greatest benefits for the advancement of the sciences from an expedition whose principal goal is to favor their progress.[33]

(2)

Mémoire par l'Académie des Sciences, pour servir aux Savans embarquées sous les orders of M. de La Pérouse

[ca. May-June 1785]

[Their report was divided into sections produced by the various scientific specialists within the Académie: Geometry, Astronomy, Mechanics, Physics, Chemistry, Mineralogy, and the two sections below germane to our study. No section was specifically entitled Natural History.]

The attention and curiosity of those who have undertaken major voyages were quite naturally drawn to the diverse varieties of the human species. Most of them seem to have been fixed upon observing and describing external characters

33. Milet-Mureau. 1797. *Le Voyage de La Pérouse*. 1: 155-156.

extracted from color, stature, conformation, and other differences of the same sort, susceptible to easy perception even to untrained eyes.

It would be desirable to extend such comparisons to the interior parts through anatomical research. Toward that end, the head bones and the hyoid bone should be procured from a cadaver of large size from every nation that appears to differ visibly from those of the temperate countries of Europe, whether in facial form or in the form of the entire head. This should enable us to acquire knowledge about the varieties found in man through comparison of the form of the bones in the head.

To obtain an even more interesting aspect of this knowledge, we should be able to compare the bodily proportions of men from different nations. . . . It is also necessary, therefore, to measure the height in a straight line from the base of the heels to the top of the head. Other dimensions to be measured very carefully include the span of stretched arms; the length of a single arm, from under the armpit to the extremity of the middle finger; the circumference of the head at the forehead; the circumference of the chest at the level of the nipples; and the circumference of the stomach at the navel.

Anatomists have found that the number of lumbar vertabrae sometimes varies from 5 to 6. In those countries where the men are notably tall, cadavers should be examined to determine whether they have 6 lumbar vertibrae.[34]

Botany: The various voyages undertaken in years past have enriched botany through the discovery of a multitude of plants hitherto unknown; and this kingdom of Nature is so fecund that we may build hopes for a new harvest from the searching by our travelers.

But it must be hoped that these searches will be especially directed toward useful things, such as a knowledge of those plants used by the inhabitants in the different places where the travelers have gone ashore, whether for food, for medicine, or relating to the arts. They should also bring back specimens and seeds of such plants, representing the useful parts, but provide us with complete descriptions. This category includes nearly all the woody plants that are used for dyeing; those used by cabinet-makers for both useful and ornamental products; and certain roots, bark, and leaves of commercial value, whose location will henceforth arouse our interest to the degree that their use becomes more familiar to us.

In general, navigators have been too dedicated to making a rich and varied collection of the seeds from exotic herbs and trees, gathered in a climate that does not vary too obviously from that of France; and whose productions, in becoming

34. Ibid. 1: 165-167.

naturalized to our climate, can serve eventually to decorate our plantings or to extend our artificial meadows.

In New Zealand, a plant in the Lily Family is cultivated known as New Zealand flax [*Phormium tenax* J. R. & G. Forst., an agave]. They use its fibers to make sail cloth, cordage, and various fabrics. Captain James Cook brought back a large supply of the plant's seeds to England, none of which came up. Transporting the roots of that plant could provide one of the handsomest presents that voyagers could make for our climates.

In France, we only have the male individual of the paper mulberry, *Morus papyrifera* L. [=Broussonetia papyrifera (L.) Vent.], which is used in China to make paper and on the island of Tahiti to make cloth. We only know the female individual of the weeping willow, *Salix babylonica* L.

The male individual of a species of dioecious strawberry named *Fragaria chilensis* [chiloensis] L. is also unknown to us. It grows naturally in Chile from where it was brought back by Amédée-François Frézier. The fruits of this plant sometimes reach the size of a chicken egg in its native soil, but are a great deal smaller from the stock cultivated in France. This difference might derive in great part from the lack of the male individual. Using plants of the hautboy strawberry [*Frageria moschata* Duchesne], as is done, may be an inadequate substitute to fertilize the female plants of *chiloensis*. If by chance the navigators find themselves in countries that produce any of the above plants, they should endeavor to bring back the missing sex of the species cited.[35]

(3)

Questions proposées par la Société de Médecine, à MM. les voyageur qui accompagnent M. de La Pérouse, lues dans le séance du 31 mai 1785.

[La Pérouse carried three naval surgeons on his two frigates, as well as one botanist with a doctorate in medicine and two additional naturalists whose medical credentials remain obscure. Entrecasteaux would be somewhat better served: he would have two surgeons on each of his two frigates, and three of his four naturalists had medical degrees.

In the preceding volume, we paid considerable attention to the gradual emergence of botany as a science, independent for the first time from medicine and herbalism. Even so, most botanists in the late 18th century still received their training as naturalists as part of the curriculum for the doctorate in medicine. After receiving that degree, they might eschew the practice of medicine in favor of botany; yet, their training had prepared them to observe human beings as a

35. Ibid. 1: 171-173.

category within natural history. Their medical training, furthermore, was more extensive and sophisticated than what the surgeons of that day received.]

I. Anatomy and Physiology

Most voyagers have written about the form and general bodily structure of the people they have observed in different countries; but it is well-known that their descriptions are generally very full of exaggeration and error. We have grounds to expect greater precision from the scientists who are accompanying M. de La Pérouse, and we beg them to take special note of the following objectives:

1. The typical structure of men and women: the large and small diameter of the head; the length of the upper and lower extremities, measured from the articulation of the arm to the extremity of the middle finger, and from the thigh to the large or the second toe; the circumference of the pelvis, the breadth of the chest and that of the shoulders; the height of the vertebral column, measured from the top of the first vertebra in the neck to the sacrum.

2. The appearance: the color of the skin and of its varying regions, and of the hair and nails.

3. The particular form of the head or the cranium: that of the face, especially the forehead, nose, eyes, ears, mouth, chin, teeth, tongue, hair, and beard.

4. Those various regions of the body that the islanders are wont to deform, whether by holes, incisions, or by the insertion of foreign objects; as well as with oils, with colors prepared from ochres or plant juices. It would be useful to describe exactly the processes used by the aborigines to give the skin indelible markings; the substances they use for that effect; how they prepare and apply them; the age or the circumstances when they practice such operations; and, in particular, the alterations or local deformities (or the effects relative to the individual as a whole) that result from them.

5. Defects, exaggeration, or different conformation of bodily parts, such as the lengthening or flattening of the forehead, the distension or narrowing of the nose, the enlargement of the mouth or ears: Are such differences an invariable character of the natural organization? or the effect of some special practices? William Dampier (English buccaneer and navigator, 1652-1715, whose published accounts of his voyages to the South Sea became famous) claimed that the inhabitants of Van Diemens's Land [Tasmania] lacked two front teeth. Is that naturally so? or because they pull them out? Are such structural formations as the Hottentot's vulva in women, the amazing elongation of the scrotum in men, or the brown spot on the backs of children observed in several regions of America of natural origin? Practically nothing has come down to us about the use of the two hands. The question of ambidexterity, or the preference of one hand over the other, has insufficiently interested naturalists. It is important, therefore, when

visiting different peoples to notice whether they use both hands equally in their work, or whether one of them is given preference; and whether the preference for the right hand, as in the refined nations, is only the effect of prejudice. It is also important to examine, among people accustomed to going nude, whether they use their feet with the same agility as their hands, and for the same purposes.

6. We have not had reliable information about the comparative strength of men. It would be well to make experiments about the loads that can be carried by inhabitants of countries where the natural constitution has not been weakened by domestication and by all the practices permitted within the refined nations; and by the distances they can cover in a given time, both walking and running.

7. The nature of the senses of sight, hearing, and of smell can provide important evidence about the vigor or weakness of those organs. Much has been said about the keenness of the aboriginal peoples' sense of smell. It would be interesting to try to verify that claim and to discover, in individuals where keenness of smell exists, whether it is at the expense of some other sense.

8. Vocal utterance, more or less distinct, as well as expressions of joy, pleasure, and pain, are important matters to be investigated.

9. The age of puberty for both men and women: Are the latter susceptible to the periodic flow in all climates? Does its abundance depend upon the climate? What is the time of its cessation? How do they conduct themselves during pregnancy? Do they give birth easily or with difficulty? and do they require assistance during the procedure? Do they tie the umbilical cord? Is that procedure accomplished before or after the expulsion of the placenta? Do they swaddle their infants, or by what means do they substitute for the swaddle? In regard to newborn infants, what practices are followed at the moment of birth such as molding the head and washing them? Do the mothers nourish them? and up to what age? Are more boys born than girls?

10. How many children die between birth and the age of puberty? In general, what is the duration of life among men in various climates?

11. The pulse beat compared to that of Europeans, which is nearly 65-70 pulsations a minute.

12. The relation of skin color to the color of bodily fluids: In men, who are more or less dark-skinned, is the spermatic liquid, the cerebral pulp, and the blood a reflection of the tint of the skin? Does skin color vary amongst the Blacks in some individuals, such as albinos or the pallid? Is such variation the result of an illness? or of a constitution altered under the influence of climate, as is believed about Blacks transported to cold countries?

13. In America, do you frequently find men whose teats contain a sufficient supply of milk to nourish children as has been said? What must we think about

the hermaphrodites of Louisiana? Does aboriginal life amongst various peoples render sexual activity periodic?

14. We do not comment here of giants or dwarfs, nor about men with tails and so on, as these alleged departures from nature have only been seen by prejudiced or ignorant travelers, or existed only in their exalted imaginations.

II. Hygiene

This branch of medicine offers travelers the most accessible observations, but is the one about which we ask the fewest questions, as it has generally been reported with greater care and detail. We shall limit ourselves to the following matters:

1. The nature of the air in different locations as measured with eudiometers: the highest and lowest temperatures in both sun and shade; its dryness, its humidity, its weight, its elasticity. Its electrical state as measured with different electrometers, in particular the one used by Horace-Bénédict de Saussure (1740-1799, Swiss alpine botanist and physicist, who published Essay sur l'hygrométrie in 1783). The divisions of the seasons; the prevailing winds, or their variations; the nature of atmospheric phenomena such as snow, hail, rain, thunder, hurricanes, and marine and terrestrial spouts; alteration of the air caused by fumes or emanations from plants by comparing (in the manner of Jan Ingen-Housz's experiments on the evolution of oxygen; the fluids that are given off by different plant parts when exposed to shadow or to sun, especially in areas known locally to be dangerous.

2. The examination of seawaters, at different heights or depths, more or less close to coasts, as well as fresh or brackish waters, for the nature of the salts they contain. The use of Torbern Olof Bergmann's reaction principles are recommended and evaporation in particular. Examination of waters drunk by islanders and the various uses they make of them; mineral waters, cold or thermal; artificial beverages, still or fermented, and the manner of preparing them, the plant or animal substances from which they are composed, and their effects. We stress the kava in particular, a liquor prepared from a root on the islands of the South Seas, to which William Anderson attributed a numbing effect and the property for dessicating and shedding scales from those islanders' skin who drink it to excess.

3. Foods: Are the inhabitants of the different countries you visit nourished by plants or animals, or by both? What food preparation is involved? Do they eat at regular hours, little or abundantly? What comparisons can be made between our vegetables and the roots, fruits, etc., that they use? What is the species of fern, containing a gelatinous substance, that feeds the inhabitants of New Zealand? Do they not have powdered foods to sustain them on trips? From what plants are they made? And what processes do they use in their preparation? Are there not to be found, through research, some plant materials unknown to the natives that could

be used for food? Should they not have some plants, analogous to sugarcane, from which a sugary substances could be drawn, but with greater ease and less expense than with the latter?

4. Habitations: their form, their space, their openings, their exposure; the soil on which they sit, the materials from which they are constructed, the degree of shelter they provide, their dryness or dampness; whether the inhabitants retire there during the night, during the whole year, or only in certain seasons; how much time do they spend in them during the day; relative to their space, in what numbers do they gather there; whether they sleep there on beds, on mats, or on the ground; whether there are some men who have no habitation and always live in the open. The clothing: forms, materials, differences.

5. The occupations of the two sexes, their work, and their physical exercises, all in respect to how they maintain or undermine the health of the people.

6. The passions, manners, the dominant character of each nation; particular usages capable of stimulating the excretion of the different fluids [humeurs] such as chewing tobacco or betel, or of smoking, using frictions, oiling the skin, hot or cold baths, dry or moist vapors; the massage method; the influence of these various means, in particular, oily ointments and tatooing, upon perspiration.

III. Illnesses

Maladies peculiar to those climates to be visited can provide important information. Cook and Anderson have mentioned, although with little detail, maladies that they observed on the Tonga Islands and the Society Islands. Among the inhabitants of the first group, they noted a deprivation of sight due to defects in the cornea; large ulcers, a serious affliction as many of the people lose their noses; a leathery, but painless, swelling of the arms and legs; and testicular tumors. William Anderson, really responsible for such observations, referred to five or six illnesses he had seen on Tahiti; but he only wrote more specifically about anasarca and about *sefaii* (painless swelling); and about the venereal disease that Captain Cook's crews had brought during two prior visits.

Because Anderson never saw a bed-ridden patient, and even though the islanders of the South Seas in general neglect to treat their diseases routinely, we urge the voyagers to pay special attention to the following points, several of which relate to diseases regarded as new in our climates:

1. Amongst the islanders, are there acute diseases or fevers? (Anderson only noted the chronic.) Are the eruptive diseases simply the most obvious to the observer? Does smallpox exist there? What about its progress and ravages? Are the people you visit acquainted with innoculation? Is that disease endemic to a particular climate? Do they experience outbreaks of contagious or epidemic diseases? Have they endured the scourge of the plague? We ask for a precise

description for all the diseases in this group, above all relative to their crises, and the similarities of their progress and nature to those of our climates.

2. Among the chronic diseases, the most frequent on the islands of the South Seas appear to be those that attack the skin. To what may be attributed the multiplicity of scurfs noted by Anderson? Are they due to oily ointments or to insect bites? Is their incidence as frequent as that of ulcers in hot countries? Those ulcers that redden the face and destroy the nose, are they not cancerous? Are the inhabitants subject to leprosy? The painless swelling of arms and legs, observed by Cook, can that be attributable to leprosy? Are they subject to disease caused by lice and the guinea worm?

3. Does venereal disease exist in those lands you will visit, whether continental or island? Does it appear to be native or has it been introduced? What remedies are used to combat it? What has been its progress on the Tonga and Hawaiian Islands where Cook brought it during his first [two] anchorages? What are its symptoms? Is it true that the islanders have no gonorrhea?

4. Is scurvy endemic in any place? What are the symptoms and its ravages, in both hot and cold countries? What remedies are used to combat it?

5. Is rickets, and its consequent deformities, known in the countries you will visit? Do nervous diseases, convulsive or spasmodic, notably epilepsy, exist there? Are children subject to particular diseases, above all to convulsions during teething?

6. Are certain men or women especially responsible for the curing of diseases? What remedies or what processes do they use? Are there any hospitals? or are any categories of the sick kept separate from society?

IV. On Materia Medica

According to William Anderson's report, it appears that only priests among the inhabitants of the South Sea islands are responsible for the care of the sick, and that they employ some herbal juices. But he says nothing about such plants, nor about the various methods by which they are put to use against skin diseases, ulcers, swellings, and anasarca. The women, according to the report, heal the consequences of childbirth by sitting on hot stones, which are enveloped in two pieces of cloth, between which a species of mustard has been placed. This treatment makes them sweat profusely. It has not been successful for the treatment of venereal diseases.

In sum, these peoples have very little knowledge about the properties of the remedies that Nature offers them. They do not even have a vomitory. It is up to the voyagers, therefore, to recognize the medicinal virtue of local plants whose taste and other physical properties should be able to bring some enlightenment, not only to the islands of the South Seas, but to any other country where they might land. Given that objective, we propose the following points:

1. Examine the taste and the odor of roots, wood, bark, leaves, flowers, fruits, and seeds of the plants within the various little-known countries and compare them to the different vegetal substances employed in Europe as medications. Do the same thing with the saps that seep from trees as well as with animal materials.

2. Examine the various remedies that are used in the hot countries where you will land to combat those diseases that afflict the inhabitants. Describe even the superstitious practices, as they are often the only medicine possessed by barbarous peoples.

3. Test the decoctions made from emollient, aromatic, or bitter plants used for those skin affections to which the islanders are susceptible.

4. Use mercury in frictions against venereal diseases that attack the inhabitants of the South Sea islands, and provide them the means to rid themselves of that terrible calamity. Above all, make note of the effects of mercury on these peoples.

5. On those islands, look for some sudorific plants [that induce sweating], which might have the antivenereal property, such as *Lobelia siphilitica* L. (*rapuntium Americanum flore dilute caeruleo*) and *Celastrus inermis* L.

6. In hot countries, try to determine whether there are plants analogous to quinine, simaruba, ipecacuanha, camphor, opium, etc., and whether the islands contain emetic or purgative plants from which some benefits could be drawn.

7. Obtain information and make observations about the properties of anacard [cashew-nut], which, in Louisiana, is said to cure madness; and on the properties of [*Sedum*] *telephium* [L.] and seagrass, which the Greenlanders prefer to cochlearia [scurvy-grass] for the cure of scurvy; and about Winter's bark [*Drimys winteri* J. R. & G. Forst.], the root from Belaage [Madagascar], from Columbo [Ceylon], and Juan Lopez's root [from the coast of Mozambique].

8. Indicate which people poison their arrows and what substances they use to achieve that effect; the nature and description of the plants from which they obtain the poisonous juices; and, in particular, the remedies they administer to counteract the destructive action. Determine, moreover, whether salt and sugar can be claimed to be antidotes for wounds made by such arrows as one could be led to believe after the experiments of Charles de La Condamine (1701-1774, chemist and mathematician, who promoted inoculation for smallpox).

9. Examine poisonous animals, snakes and fish especially, to try to discover what produces this dangerous property, and what means can be taken to forestall any harm.

10. Take care to collect those remedies, whether internal or external, that are

claimed to be specifics for diseases by different peoples. Describe the nature of such specifics, the way they are prepared, the way they are used, the doses, their effects, and at what point in the illness they are employed. This is the way the Peruvians taught us to understand the properties of quinine.

11. Finally, assemble a special herbarium, separate from the botanical herbarium, for those plants or plant parts that are used for food, as remedies, or as poisons in all the countries where you go ashore.

V. Surgery

Anderson remarked that surgery is a bit more advanced than medicine, as one ought to expect among peoples little subject to diseases, but exposed, as is mankind in general, to external accidents. Cook mentioned a woman on New Zealand who practiced the vocation of oculist. She was treating a child's eyes with two small wooden probes, rubbing on the organs to a point of drawing blood.

It appears that the natives of the Tonga Islands do not fear wounds as they make them freely on their heads as a sign of grief. They cut off a little finger with a stone ax when ill, or one of the phalanges of that finger when a local chief is sick. You will see many islanders with either one or both little fingers missing. No doubt, this practice emerges from superstition. They also make incisions on various parts of the body, on the legs in particular. Beyond which Anderson remarked that they are very poor surgeons. He had seen a badly-cut arm with a dislocation that had not been set after several months.

Finally, the men on Tahiti, called *tahous*, who are employed to remove the foreskin on infant boys. They accomplish this with a single incision made with a shark's tooth. They treat the subsequent swelling to that part by applying hot stones to it.

It appears to us that the important facts to be gathered about surgery can be reduced to the following points:

1. Dislocations, fractures, hernias, and surgical ailments in general: Are they very common among men who live in the state of Nature?

2. What means do they employ to heal these different ailments?

3. Do they have any special instruments? What is their form, material, use? Buy some of them to make a collection.

4. Are circumcision and infibulation practiced? What are the procedures?

5. Are there any men or women especially charged to treat a given surgical ailment, such as with the eyes, the ears, teeth, skin, or childbirth?

6. What is the form and nature of weapons used in combat, that of the resulting wounds, and the manner in which they are treated and healed?

At the Louvre, 31 May 1785. Signed Mauduyt, Vicq-d'Azyr, Fourcroy, Thouret.[36]

(4)

Within the general instructions prepared for La Pérouse, and ultimately for Entrecasteaux, were provisions seemingly general in application but, in fact, pertinent in particular to the naturalists on board. At ports of call, they would be the personnel most frequently ashore to carry on scientific research as intensely as possible, therefore most likely to encounter the aboriginal peoples. Moreover, they would make substantial collections and keep written records of such people and their culture. In both matters, Fleurieu had defined proper conduct in a precise manner that expressed the moral character of the entire enterprise.

Both commanders were provided reports from prior overseas expeditions with the explanation that His Majesty did not doubt that, informed by that reading, they would be moved to imitate the good conduct of some of those navigators and to avoid the faults of the remainder. They were to take pains, upon arrival at a new country, to gain the friendship of the principal chiefs by indications of good will and with presents. All honorable means were to be employed to promote bonds with the local natives.

Even while endeavoring to determine what merchandise of European origin the natives valued the most as a foundation of an equitable exchange of goods, they should remain aware of the necessity to take precautions to maintain an appearance of strength without actually using force. No matter how warm the welcome might seem from aborigines, the visitors must be perceived as in a state of defense so as not to encourage the natives to attempt a surprise. Fleurieu recommended in some earnest detail procedures to facilitate a successful trade with aborigines that would enable the French to obtain the fresh water, food, and fuel the port of call could provide, but without engendering conflict.

All personnel on board, he added, must be made to understand the requirement to live on good terms with natives and to seek their friendship through correct proceedings and considerations. It must be understood that it is forbidden, with the threat of the most rigorous penalties, ever to use force to take anything the inhabitants refuse to give up willingly. All peoples visited in the course of the voyages were to be treated with great gentleness and humanity.

Within that context, Fleurieu added three final paragraphs: "The commander will devote himself, with energy and interest, to promote the improvement of the

36. Ibid. 1: 180-196. Dr. Félix Vicq-d'Azur, co-founder of the Société de medicine in 1776, of which he was the permanent secretary, was a leading French anatomist and a promoter of comparative anatomy.

aborigines' condition by introducing the vegetables, fruits, and useful trees from Europe to their countries; by teaching them the method for planting and cultivating them; by making them understand the use they should make of these presents, the objective of which is to increase the necessary productions from their soil for peoples who draw nearly all their nourishment from the ground.

"If, under extreme circumstances, which prudence requires must be anticipated during such a long voyage, the commander should ever find himself obliged to make use of the superiority of his armament to obtain vital necessities such as food, wood, or water, despite the opposition of the aborigines, he will use force with the greatest moderation, and will punish with extreme rigor any of his people who go beyond his orders. In all other instances, if he cannot secure the friendship of the aborigines by kind treatment, he must seek to restrain them through fear or threats. But he will have recourse to arms only as a final extremity: only for his defense or on occasions when the security of his launches and the lives of the French, whose conservation has been confided to him, should be unquestionably compromised.

"If the expedition can be terminated without having cost the life of a single man, His Majesty will regard that result as one of the most commendable of its successes."[37]

As these two expeditions were planned and financed by the state, and carried out on ships of the royal navy; and as all of the participating naturalists served as employees of the state, the government logically assumed that all drawings made during the voyage, all boxes containing natural curiosities, as well as the descriptions made of them, and the collections of astronomical observations, would be turned over to the commander of the expedition at the end of the voyage. The instructions were explicit: No scientist, no artist, would be able to reserve for himself, or for others, any of the natural history objects, or any other item, that the commander will have judged to merit inclusion in the collection destined for His Majesty.

"Before reaching the port of Brest at the end of the voyage, or before reaching the Cape of Good Hope in the event of putting into that port, the commander will require all officers and marine guards, the astronomers, scientists, and artists, the pilots and any other personnel on the two frigates who have kept travel journals, to turn them over to him. He will enjoin them to maintain an absolute silence on the objectives of the voyage and about those discoveries that will have been made,

37. Ibid. 1: 50-54.

even requiring their parole. In addition, he will assure them that their journals and papers will be returned to them."[38]

Whether such instructions were understood by the botanists from the outset of the two voyages remains obscure. They would never apply to the La Pérouse expedition, to be sure, as it would be lost at sea with the bulk of its collections. But among the wretched wrangles that would become a dismal feature of the premature termination of the Entrecasteaux expedition on Java would be a dispute between naval commander and botanists over possession of botanical specimens and journals. It would only be settled by the benevolent intervention of Sir Joseph Banks in favor of botany.

38. Ibid. 1: 49-50.

Portrait of Jean François de Galaup de La Pérouse.
© 2000, Hunt Institute, CMU, Pittsburgh, PA 15213, U.S.A.

III
The French Gardener Overseas

There are two principal reasons for giving time and space to the detailed instructions drawn up by André Thouin for the gardener who accompanied La Pérouse, which would serve again to guide the gardener with Entrecasteaux. The gardener was officially and socially subordinate to the botanist on board, the one a savant, the other a skillful workman, albeit a well-trained horticulturalist. Yet, he was also indispensable to the botanist for the management of live plant materials that were wanted for scientific observation in Europe; and he had the additional, humanitarian responsibility for what Alan Frost has called *the antipodean exchange*, the transmission of *useful plants* between Europe and the exotic world.

Frost has effectively described how the British in the 18th century developed a viable network of trading stations overseas by planting gardens and fields that could sustain settlement while providing frequent refreshment outposts for trading vessels on long voyages. The conception was suitable for a maritime nation, and the French had nothing comparable, as he reveals by describing the astonishment of the French when the Baudin expedition put into Sydney in 1802.[39] The French neglect of such development had already been apparent from the instructions prepared by Thouin, which were a clear reflection of the humanitarianism of the French Enlightenment and bore no resemblance to official British practice.

Historians have always known of the vast importation of exotic and unknown plants to Europe in the 18th century, but generally have remained unaware of the laborious procedures required to preserve the viability of incoming and outgoing plant material. Specimens intended for herbaria, to be sure, were put into presses for drying and preservation in the conventional manner. But the botanical mission had even broader dimensions: the selection of useful plants and trees for French gardens and nurseries where they could be acclimated for reforestation or ornamentation; and the donation of French fruits and vegetables to peoples overseas for their nutritional benefit.

If, in the outcome, the French realized greater profit from the exchange than did the exotic peoples, their greater success should be attributed to their more

39. Alan Frost. 1996. "The Antipodean Exchange: European Horticulture and Imperial Designs." Miller & Reill. *Visions of Empire*. Cambridge: Cambridge University Press, pp. 58-79.

advanced technical knowledge and their conception of progress. Whereas, the absence of what we would today call ***an infrastructure***, whether institutional or intellectual, can defeat the best intentions of reformers. The inequitable outcome has led to the quick-and-easy charge of deliberate exploitation, when the very opposite was intended. Thouin's instructions expose, what is more, the careful preparations for a fruitful exchange.

Of all the presents the munificence of the king means to give to the inhabitants of newly-discovered countries, Thouin wrote, plants useful for the sustenance of mankind are unquestionably the ones that will bring them the most durable benefits and are most likely to increase their happiness. Those vegetable-garden plants and fruits that require no preparation must take first place, while those that need only dry cooking to become edible must be given second rank. The latter alone should be given to people who have no vessels appropriate to cook foods in liquids. Furthermore, the seeds of vegetable varieties that require a painstaking, scrupulous cultivation should not be given. Such seeds, left on their own in climates so different from that of their origin, or more or less abandoned to a crude cultivation, soon degenerate into their primitive species.

To make our vegetable shipments in the form of seeds, he continued, is undoubtedly the least expensive way to acquire them, the easiest way to transport them, and the surest way to multiply productions from one climate into another. The most scrupulous attention must be paid to acquire seeds from the most recent harvest that are well-ripened and completely clean. Any found to be wrinkled or bitten by insects should be carefully excluded, not only as infertile, but as noxious to the conservation of the remainder. Once assembled, the seeds must be divided into two groups. The first group includes only those seeds that merely require shelter from air and humidity for preservation. The second group includes those that must be covered with sand or soil to preserve their germinative property, such as our fruit-tree kernels or the seeds in several plants in the *Umbelliferae* (Carrot Family), etc. Those in the first group need to be enclosed in bags of brown paper, and then be placed in tin boxes that should be made fast, as hermetically sealed as possible. The second group should be put in tin boxes in layers with their soil or sand, and then very tightly sealed.

The tin boxes, after proper sealing, must be enclosed in stout packing-cases, which are then wrapped in oilskin. They must be stored in that part of the ship least accessible to humidity and must be sheltered from extreme heat and deep cold, and left unopened until the moment when the seeds are to be planted. It is improbable that the seeds will have a single destination. To avoid exposing seeds meant for different locations, it would be convenient to divide the totality of the assortment into at least four sections, so that only one packing-case will need to be opened for a given planting, avoiding unnecessary openings that could be prejudicial to the seeds.

Order being absolutely indispensable, the gardener will take care to write the name of the seed on each packet and to inscribe it on a register as the packets are successively put in the tin boxes; and will number those boxes correspondingly in his register, placing them in order by number in the large packing-cases.

When going ashore in a place where it is proposed to make some plantings, the gardener's first concern must be to determine the regional temperatures and to note whether the products of the soil, the annual plants in particular, are growing, vegetating, or withering. Such observations will guide him in a choice of seeds suitable to the climate and suggest the exposures most favorable to the planting. In very cold countries, if you should arrive in autumn or winter, abandon any idea of sowing the seeds of annuals. They will either not germinate or will be destroyed by the first frosts. You can risk some seeds from woody plants such as pips from apples or grapes, or the kernels from different fruits. Those seeds will not come up before spring and will preserve themselves despite the cold. If you arrive in the spring or summer, then all seeds likely to acclimate themselves may be planted, taking care to choose the soil and exposure suitable to each variety.

In many hot countries, dryness can be the factor most harmful to the multiplication of plants. It is best to select moist soils, stream banks, and the low terrain near the sea. Shady places ought to be preferred for the establishment of gardens.

Once the places for planting have been selected, the gardener will have to work the soils in preparation for seeding. He will then plant his seeds and watch over their development for so long as the ships remain. If he can inspire a love for these plantings in a few of the local natives, getting them to understand the value of those products that are the objectives, he will doubly fulfill the benevolent goal that we hope for his mission.

To avoid overlooking any of the ways that can make his voyage useful and gratifying, the gardener must maintain an exact record of all his activities: the date of his plantings, their success, the development of the plants and their outcome. When he has the opportunity to observe that cycle, we shall be provided comparative data that can improve our own cultivation.

As a number of plants exist that could be useful to the inhabitants of places visited, but whose seeds do not possess the property of perpetuating our satisfactory varieties such as the greater part of our fruit trees (the products of a lengthy horticultural development), it is advisable to carry several live individuals of each of those trees. Despite that, you should not fail to carry an abundant supply of seed from such trees. They will provide wild fruits such as those that fed our ancestors, useful to nourish people even less civilized than they were, providing them a foundation of wealth from which their improved skills will henceforth be able to extract the greatest benefit.

The Transport of Live Plants

As it is not now possible to dig up plants in the open terrain as they are in full leaf, you must buy them in pots from tradesmen in Paris and take them to the point of embarkation. The transport of live trees can only be made in packing-cases where they can vegetate during the voyage. Use a tub 40" long, 20" wide, and 20" deep, pierced at the bottom with a dozen holes for the run-off of excess water. Over the top, there should be a triangular framework upon which can be fitted wire netting, some skylights, and some shutters to provide for free circulation of air, to increase warmth when there is need for it, or to shelter from the cold.

(Thouin's list of recommended fruit-tree species was appended to his memoir.) It is advisable to obtain only young individuals, branched along the full length of the stalk. Take care that they are well-shaped, vigorous, and that the grafting was done as close to the root as possible.

You pack them in the following manner: At the bottom of the tub, over the holes cut for drainage, you place plaster stones to prevent the soil from washing away during irrigation. Over them put down a bed of good soil 3" thick and tamp down lightly. The first layer of trees goes over that bed, selected from the largest and, in particular, from along those like figs, grapevines, and cherries, which tolerate having their stalks buried a bit deeply. After removal from the pots, the balls of those trees must be placed as close to each other as possible, the spaces left between them to be filled with heath-compost, compressed as much as possible so that the first layer forms a solid mass. A 2"-bed of heath-compost is then added to cover the first layer.

The second layer must similarly be arranged with earthen balls against each other, the tallest stalks in the middle, the shortest stalks, by gradation, on the edges. All spaces must be filled with heath-compost without disturbing the stalks in the lower bed. Assuming that the stalks had three or four buds when taken from the ground, this packing should suffice for their preservation. Consolidate the entire package, finally, by knocking the packing-case against the ground or by pressing with a fist; so that no empty spaces remain, and no jolts from carriages or the rolling of the ship can provoke any displacement.

The contents of the packing-case may then be well-watered. Within a few days thereafter, it may be taken to Brest by wagoners. The lateral shutters should be closed during that trip to reduce the loss of moisture, but the two small openings on the extremities must remain open for circulation of air so that the plants do not rot. Upon arrival in Brest, the gardener should lift the wire netting in order to clip any shoots put out by the trees.

Management of the packing-cases during the voyage will be confined to irrigations when necessary and to protect the trees from extremes of hot and cold: whether by covering them by canvas during the day, and providing them the most air possible during the nights; or by putting them below deck when on cold stretches of ocean. It will be helpful from time to time to use a pruning-knife on overly vigorous plants that, otherwise, might harm neighboring plants.

Once the trees reach their destination, they will be removed as carefully as possible with their earthen balls. They should be planted where the exposure and the nature of the soil is most suitable for each of them. During his sojourn, the gardener will watch over their preservation. The gardener will only remove from the packing-cases the stock meant for that particular place. The vacated space should be refilled with such local productions that he believes ought to be useful in Europe. The gardener, finally, must strive to make the local inhabitants understand that these trees are presents, and that they must care for them conscientiously in order to benefit from their usefulness.

On the Collecting of Plants that can be Useful in Europe and their Preservation During the Voyage

Seeds are best collected in their full maturity. But, as the brief sojourn on those islands where you go ashore may not allow the time to wait for full maturity, seeds may still be collected by taking indispensable safeguards.

Herbaceous plants, whose seeds are found to have reached only two-thirds or three-quarters of their maturity, should be pulled up by the roots, tied in bundles and taken aboard ship, where they must be hung in a place sheltered from sun and moisture. A portion of seeds from these plants will mature without question within six or eight days, at which point they may be collected.

In an instance where you are determined to collect seeds from an interesting plant, but find the seeds barely set, the project need not be abandoned. Such plants may be dug up, retaining a ball of soil around the roots, and planted in baskets. During the first days thereafter, the baskets should be covered with matting, the plants inside watered morning and night. You will discover that maturation of the seeds will occur slowly during the voyage, and you will not have had to forgo an often unique opportunity to procure a valuable plant for Europe.

When you have the good fortune to encounter fully matured seeds, it remains important to collect them in a prescribed manner for their preservation. Not only should you avoid shelling them, but, on the contrary, they should be collected with their envelopes and their peduncles. Whether borne within pods, siliques, or capsules, they should remain within their fruits. It will even be necessary to tie them closed so that they do not open during the voyage. This applies equally to cones and, in general, to all dry fruits. Tiny seeds growing in panicles, spikes, whorls, and corymbs should be collected in their entirety with stems 5-6" long that can be wound around in a manner to prevent exterior air from reaching the embryo in the seeds.

During a voyage so long in very different latitudes, seed preservation requires special precautions. While it is unquestionable that seeds remaining within their envelopes will preserve better than those that are shelled, they must be quite dry, along with all parts that accompany them. Attention must be paid to rid them of insects, along with insect eggs that could hatch during the voyage, as the seeds could be eaten unperceived. Several hours exposure to a direct sun should suffice to produce the initial effect. If imperceptible insects or their eggs are suspected to be within the seed envelopes, by taking the precaution of putting them under a bell jar within which some sulphur has been lighted, the fumes from that substance will kill them in a few minutes.

Assured that the seeds contain neither superfluous moisture nor insects, you may wrap them, species by species, in squares of heavy paper that will not be sealed. Each packet should be given a number identical to that of the plant or tree specimen from which that seed was collected. Next, these various packets are to be stored in a tin box by squeezing them together as tightly as possible, not only to economize on space, but to protect the contents from the lurching and rolling that could cause rubbing of unsecured seeds against each other and self-destruction. The filled box will be covered by its lid and sealed shut as hermetically as possible. You should immediately write on the box what it contains, for example, seeds collected from this date to that date in this place. When several similar boxes have been assembled, they are then to be packed in a chest of heavy wood, to be covered by a waxed canvas, upon which information comparable to that above will be written.

Bare seeds, the size of a hazelnut or bigger, will require a different preparation. Immediately after their collection, it is customary to leave them exposed to the open air in a sheltered place in favorable weather so that excessive moisture within them will evaporate and complete their maturity. After which you inspect these seeds, rejecting any found to be deformed, aborted, or bitten by insects. Then, in a tin box with sufficient capacity to hold double the bulk of the seeds, you will put a layer of soil one finger thick over the bottom, upon which you will put a layer of seeds, separated from each other by a very small space. These seeds are then covered by about a half-inch of soil, over which you put a second layer of seeds. And so on until you reach about one finger from the top of the box, which is then filled with soil, covered tightly with a lid, and sealed.

The soil used in this procedure must be neither too dry nor too moist. If too dry, it will absorb the moisture necessary for the preservation of the seeds. If too moist, rotting will result. Thus, success in this kind of packing depends upon finding the middle between two extremes, relative to the nature of the seeds. The label you will put on the sealed box must always correspond to the gardener's herbarium and journal. The importance of this precaution is obvious.

Bare seeds, pea-size or less, can be put together with soil without any need for regular layers, but otherwise arranged like the preceding seeds.

Seeds enclosed in fleshy calyces, whether berries or pulpy fruits like figs, gooseberries, apples, peaches, etc., must be removed when those fruits begin to rot, indication that the seeds have reached full maturity. You then lay them out in open air, after which they can be packed, with soil in tin containers, as above.

All those boxes that we have recommended be sealed, once they are filled, must not be opened under any circumstances until the moment comes in France when it is time to plant the seeds. They require no further care during the voyage other than being placed on board ship in the place least exposed to variations in the atmosphere. They must be protected from excessive moisture and from great dryness in particular.

Given present knowledge, there are some seeds whose germination seems impossible to retard, such as those of the palms, the myrtles, the madders; and, in general, those whose seeds are filled with a corniculate body, and which have only one very small embryo. These families produce numerous fine trees, for the most part useful. The lack of success with seeds from these trees, brought to us after considerable precautions had been taken, would seem to prove the impossibility of success, suggesting different means are in order.

We believe it could be feasible to plant such seeds following their collection. This would require use of a tub whose size would depend upon the number of seeds you want to accommodate, but would provide a depth of at least 20". The tub should be filled with a substantial light soil, procured at the moment of need from a place fertile in plants. The seeds may be planted quite close to one another, but at varying depths: the largest, like the coconut palm from the Maldives, at a depth of 8", to the finest seeds lightly covered. A space of 2" between the soil and the upper edge of the tub should be left for a bed of moss, which will be held in place by nailing 4 or 5 cross pieces to the sides of the tub.

Once these seeds are within the tub, a row of hoops needs to be fastened over the top to support a wire grill cover that will protect the seeds from damage by rats and domestic animals on board the ship. The culture of these seeds will require irrigation to maintain a degree of moisture favorable to germination; preserving them from direct sun by covering them during the day with a large canvas; and above all by protecting them from the cold on the open ocean when there is danger of frost by moving them into the most sheltered place on the ship. And, finally, by carefully hoeing the ravenous herbs that could be harmful to delicate new plants.

Up to this point, we have presumed that the gardener will find seeds either mature or nearly mature at places visited. But it could happen that he could meet neither one nor the other. Finding himself also unable to remove living plants, he could be in the sad situation of having to abandon that place without being able to show Europe a single example of its productions. But there remains to him one last resource even in the least favorable circumstances.

Everyone knows that plant seeds are shed as they become mature. A great portion of them are picked up by water and carried into low places, or are carried by wind to the edges of woodlands. By sweeping up the debris from such places with a broom, you can be sure of removing many seeds of native plants along with the local soil. Such soils, enclosed in boxes after having been suitably dried, will preserve the seeds until you reach Europe. We have had proof of that from a shipment made to us by J.-B.-Christophe Fusée-Aublet from Cayenne. This traveler put about 60 packing-cases on board ship, filled with valuable trees and plants from the colony. None of the trees survived the trip. But the soils in which they had been planted, having been spread over a large surface of beds covered with frames, produced a great number of plants, some of which still survive in our gardens. Therefore, this method may be used in confidence of success. In certain instances, it may be the only one able to procure some plant species.

In terminating this section on the shipment of seeds, we must endeavor to establish those priorities that will guide their collection. Without any doubt, the trees and plants likely to be useful in Europe for the nourishment of the population must be given top rank, such as that species of fern whose root serves as a food for the peoples of New Zealand. Those plants useful to the arts should occupy the second rank; those plants that can be used for the ornamentation of our gardens, the third rank; and the fourth rank should include plants only appropriate for botanical gardens.

In the collection of seeds of whatever kind, the climate in which they are found does suggest an additional priority. In those countries where the temperature is analogous to that of Europe, you risk nothing by collecting in great abundance. The usefulness of those seeds will be easy to determine before being planted in open ground, and their quantity will provide the means to distribute them widely in the various provinces of France. Those from hotter countries should be collected in reduced quantity, because, as flatbeds, cold frames, and hothouses are required for cultivation, one can only preserve a small number of them unless they are to be passed on to our colonies in the Antilles or in India.

Finally, try to collect a larger quantity of the seeds of each species during the last years of the voyage than during the first years. Is is highly probable that, despite taking all possible care, a portion of the seeds collected early in the voyage will lose fertility before the ships return to Europe. The collections from the last years of the voyage will be infinitely better.

During the voyage, if you foresee a likely anchorage at some European outpost, prepare in advance a supply of small packets of the seeds from species previously collected; to which you may add duplicate herbarium specimens whose numbers must be identical to those put on those seed-bags you will be taking to Europe yourself. Such an advance shipment made from a European outpost may be put in tin boxes, enclosed within packing-cases, then wrapped in waxed canvas, and addressed to the Maréchal de Castries for the king's gardens.

On the Transport of Bulbs, Onions, Fleshy Roots of Perennials, and their Culture During the Voyage

If you find such plants when their fructification has been completed and their green parts withered, that will be the most favorable season to take them from the ground. They will subsequently require no other care than to be cleaned for removal of any outer skins that could retain moisture and contribute to rotting the onions. If exposed to the sun for a few days, they will dry off, and you will be able to enclose them in boxes, layer by layer, with sand that is fine and very dry.

But, if you find such plants in full vegetation, it then becomes necessary to retain a ball of soil around the base when removing them from the soil. Plant them in baskets, cultivating them therein until the time when, the foliage having withered, you can remove them from soil without risk. They can be preserved by following the procedure in the preceding paragraph.

As bulbs cannot be delayed in their vegetation, when the time comes that they customarily begin to grow, they will do so no matter what you do. Therefore, the gardener must remember to examine the boxes in which he has packed them from time to time. When leaves begin to show, he should prepare new boxes for their replanting, 8" in depth, 1' at the most. They should be filled with a light soil, sandy but substantial, a soil the gardener has already perceived to be especially productive of plants. The onions, or bulbs, can be planted a half-inch from one another, at a depth of 1-4", depending upon their size. When the planting is completed, there should be about 2" of space between the soil and the top of the box to accommodate a bed of moss or, that unavailable, dried vegetation. Then nail a hatched grating over the top to protect the contents from the rolling of the ship, over which you must fasten a framework of hoops and a wire grating.

During the vegetation of these bulbs, their culture will consist of light waterings, weeding, and care to avoid too much sun, overly abundant rain, and, above all, cold.

When the vegetation of bulbous plants is complete, you may then deprive them of water entirely to accelerate the withering of the foliage and by exposing them to direct sunlight. Thereafter, you may remove the bulbs from the soil and enclose them again in their original boxes after having followed the procedures indicated above for their preservation. These procedures may be repeated for each year that the voyage lasts.

In order not to lose the history of the vegetation of these bulbs given their multiple displacements, it would be appropriate to attach a lead number to each one with a wire, the number matching the one in the gardener's journal.

On the Choice of Live Trees that You Want to Bring Back to Europe, and Their Culture During the Voyage

Only a few plants, cultivated for 3 or 4 years in packing-boxes, and experiencing frequent changes in temperature, can resist so many adversities, no matter the assiduous precautions that we shall sketch out in detail. You must rigorously limit yourself to collecting only essential items in this manner when the collection of seeds appears to be impossible.

Only very young individuals should be selected. Those coming directly from seed are preferable to those growing from root-suckers. They should be quite healthy and vigorous, their stalks having nearly the thickness of a thumb at base, and be branched from their roots if possible. Dig them up carefully without breaking or scraping the roots.

The packing-case, made of solid wood, should be set level on lath braces, which will raise the case several inches above ground, and be placed in a spot sheltered from the sun. In the bottom, and over the drainage holes you have made, put either seashells or broken stones to facilitate water drainage. Next, put down a bed of light, but substantial, soil to a depth of 2-3". If the trees and shrubs you mean to bring back have bare roots, lay them flat one against another, beginning with those with the most voluminous roots and ending with those least well provided, as tightly as possible to save space. Finally, using a fine, very dry soil that can penetrate and fill all empty spaces around the roots, you must bury the plants completely, taking care that you pour the soil on the roots and compress it. You will then water the case on several occasions until water is seen to drain from the holes. It would be well to cover the surface with a layer of moss several inches thick and put on a wooden grill to hold the contents in place.

If the plants are not bareroot but have dirt balls around their roots, probability of survival would be greater. In this event, the earthen balls should be packed closely together, upright in the packing-case. To utilize the space lost by their volume, cuttings from trees that propagate in that manner, such as mangroves, figs, and other spongy trees, can be planted among the earthen balls.

These packing-cases, bearing shrubbery, may then to taken on board ships. Their culture must be the same as is practiced in our hothouses: daily watering according to need and the warmth of the area through which you are passing. It is far better to sin through insufficience than excess. These irrigations must be made with fresh water, sea water being harmful to nearly all plants. In warm latitudes, you must water morning and evening with a perforated watering-can, in the manner of a light rain that washes the leaves and stalks before the soil soaks it up. In cold countries, you must only water because of pressing need, choosing the warmest time of the day and pouring the water only at the base of the plants that need it.

The gardener must examine his shrubbery everyday to clear away any dead leaves or insects that could be harmful. He must trim away overly vigorous shoots; he must provide shelter from the cold, extreme heat, dryness, and excessive moisture. Above all, on the open seas where it may not be possible to leave his shrubbery in the open air, he must aerate the packing-cases from time to time by opening the two small cross pieces on the ends. Otherwise, the plants will droop, mildew, and ultimately perish.

Succulents, by nature fleshy plants, such as prickly pears, the cereus, aloes, euphorbias, the woody purslanes, the ice plants, etc., may be brought back alive in the same manner as the shrubbery; but the two should not be packed together as succulents require a different culture, thus separate packing-cases. They are to be packed very close to one another in very compact soil. The packing-case should have 6-8" of it to receive them. Instead of moss, you should put in a layer of long straw or very dry hay over the soil, firmly held in place by a wooden grill. Once planted, you will water them quite abundantly to consolidate the soil around the roots. Thereafter, these plants must only be watered when they greatly need it. They must be given air as much as possible and, above all, be protected from moisture and cold.

These are nearly all the essential precautions you must take for success with live plants. The gardener's own intelligence will contribute an infinite number of petty details that have not been anticipated here. But we believe that this method be used in order to enrich Europe with foreign productions beginning with the year the ships return home.[40]

List of Tools and Utensils Required by the Gardener During his Voyage

[The Navy alotted Thouin a budget of 2,380 livres to cover the shipment of his botanical merchandise, to provide an assortment of boxes for the collected plants, and for 50 reams of brown paper for drying plant specimens. An additional special fund of 400 livres was designated for the purchase of garden-plant seeds.][41]

1. 24 tin boxes of different sizes, ranging from 10"x 8"x 6", to 20"x 16"x 12". These boxes are meant to contain a portion of the seeds to be carried from Europe, and they will serve on the return voyage to carry the seeds that are found during the trip.
2. 2 watering-cans, one with a perforated head, the other with a spout, for the cultivation of live plants on board ship, and for the beginning culture of the seeds to be planted in different locations.
3. 4 pruning knives of different sizes for use in cultures and collecting.
4. Two grafting knives for the same use.
5. 2 mercury thermometers, the Réaumur scale, for placement in the packing-cases of live plants to guide the gardener in their culture.
6. The 10 punches necessary to stamp the numbers appropriate to mark trees and plants that will be carried from Europe and those to be brought back.
7. 60 pounds of the thinnest lead sheets for making numbers.
8. One 6' staff, foot-long sections marked along its length, which, at the lower end, will have a socket into which a spade can be screwed in order to dig up plants with their roots; and another socket on the upper end for placement of a small pruning hook to reach tree branches beyond hand's reach.
9. 2 2-pronged pickaxes, one flattened, one sharp, suitable for deep digging necessary to plant trees or to remove those you want to bring back to Europe.
10. 2 spades for the same purpose.
11. 2 handsaws, closing as a knife, meant for sawing branches from which it is otherwise impossible to gather seeds.
12. 3 tin boxes, 16"x 10"x 6", divided on the interior into several compartments with hinged openings, and having rings for fastening a shoulder-strap, for

40. "Mémoire pour diriger le jardinier dans les travaux de son voyage autour du monde: par M. Thouin, premier Jardinier du Jardin des plants." Milet-Mureau. 1791. *Le Voyage de La Pérouse.* I: 205-232.
41. Ibid. 1: 245, 249.

when you will go out to collect seeds and plants for the herbarium. [Not called a *vasculum* until the 19th century.]

13. 6 additional tin boxes, 16"x 11" and a foot deep in which to put dried plants whose number should correspond to those placed on the seeds that will be collected.
14. 6 reams of large, strong brown paper, not sized, for preparing plants meant for the herbarium.
15. 4 reams of large white paper for mounting dried plants.
16. 4 notebooks in-quarto of white paper appropriate for the gardener's journal, and for the lists of the productions that he will be carrying as well as those he will be bringing back from his voyage.
17. A large writing-desk provided with 2 penknives, 12 pencils, and a stylet for dissecting seeds.
18. A double-lensed loupe.
19. Duhamel du Monceau, *Elémens d'agriculture*, 2 vols. [1754].
20. Duhamel du Monceau, *La Physique des arbres*, 2 vols. [1758].

These two works are meant both for the instruction and the diversion of the gardener during such a long voyage.

List of the Seeds that must be Acquired for Planting in the Different Places to be Visited.

1.

Items that can be eaten without preparation.

Fruits from Trees

Pips from	Apples	6 bushels
	Pears	6 idem
	Grapes	8 idem
	Gooseberries	8 pints
Stones from	Peaches	2 bushels
	Apricots	1 idem
	Plums	1 idem
	Cherries	1 peck
	Almonds	2 bushels
	Walnuts	2 idem

Fruits from Herbs.

Seeds of	Melons (various species)	6 pints
	Watermelon, red and white fleshed	4 idem
	Artichokes, white and violet	4 idem
	Malaguette pepper [*Amonum granum paradisi* L]	1 idem

Vegetables

Seeds of	Celery, different varieties	1 bushel
	Chervil	1 peck
	Garden cress	1 bushel
	Parsley, different varieties	4 pints
	Yellow purslane	1 pint
	Common sorrel	idem
	Head lettuce	1 peck
	Romaine	idem
	Curly lettuce	1 bushel
	Wild chicory	idem

Roots

Red and white onions	1 bushel
Turnip rapes [*Brassica napus* L.]	2 bushels
Turnips [*Brassica rapa* L.]	6 pints
Red and white radishes	2 pints
Garlic	1 pint
Shallots	1 pint

Cereal Seeds

To be acquired at Brest	Wheat, different species	8 bushels
	Maise, different varieties	4 bushels
	Buckwheat	idem
	Piedmontese rice	idem
	Barley, different species	idem
	Oats, different varieties	2 bushels
	Rye	4 bushels

3.

Products that are edible only when cooked in a liquid and which are only suitable for peoples who have utensils in which to cook them.

Peas, different species	6 bushels
Beans, different species	idem
Broad beans [*Vicia faba* L.]	3 bushels
Lentils, large species [*Lens culinaris* Medicus]	2 bushels
Chickpeas, red and white [*Cicer arietinum* L]	1 bushel
Lupine [*Lupinus albus* L.]	2 pints
Vetch [*Lens nigricans* (Bieb.) Godron]	idem
Fenugreek [*Trigonella foenumgraecum* L]	1 pint
White mustard [*Sinapsis alba* L]	idem
Eggplant [*Solanum melongena* L]	1/2 pint
White cabbage	1 bushel
Red cabbage	idem
Pumpkin	idem
Cucumber	1 pint
Winter squash	idem
Calabash [gourd] [*Crescentia cujete* L]	idem
Orach [mountain spinach] [*Atriplex hortensis* L]	1 bushel
Swiss chard [*Crambe maritima* L]	1 peck
Tobacco	1/2 cup

Note: It would be well to divide this assortment of seeds into four equal parts, each one to be enclosed in a packing-case that will remain unopened until the moment of planting them; thus to cope with the difficulty of exposing seeds that are not meant for use for several months or even a year after their collection.

List of Plants that Must be Transported Alive
(Fruit trees and shrubs)

1—Red winter queening apple
1—White apple
2—Pippin apples
2—Lady-apples
1—English butter-pear
2—Bon chrétien winter pears
2—Passe-crassane winter pears
2—Saint-Germain sweet pears
4—White Chasselas grapevines [Le Mâconnais]
4—Muscat grapevines [Muscatel]
2—Corinth grapes [Currants] [*Vitis vinifera* L]
2—Grosse-mignonne peaches [an espaliered variety]
1—Nectarine
2—Greengage plums
1—Mirabelle plum
2—Damson plums
2—Common apricots
2—Apricot-peaches
3—Green figs
2—Angelica figs
2—Purple figs
2—Montmorency cherries
2—Heart cherries
2—Whiteheart cherries
2—Olives
2—Portuguese quinces
1—Black mulberry
2—Cultivated chestnuts
1—English walnut
1—Soft-shell almond
2—Maltese raspberries [probably *Rubus idaeus* L]

(VEGETABLES)

To be taken from Brest	Potatoes, different varieties Jerusalum artichokes Garlic Shallots
To be taken from the Cape Verde Islands, the Cape of Good Hope, or North America	Sweet potatoes Yams

(ORNAMENTAL SHRUBS)

Summer rose bushes
Lilacs
Tuberose [*Polianthea tuberosa* L][42]

Botany books provided for the Voyage

(AMONG A FAIRLY SUBSTANTIAL SCIENTIFIC LIBRARY)

Valmont de Bomare. *Dictionnaire d'histoire naturelle* (1764)
Mustel (an agronome). *Traité théorique et practique de la végétation* (1781-1784)
Linnaeus. *Philosophia botanica* (1736)
Genera Plantarum (1737)
Systema Naturae (1735)
Systema Vegetabilium (1774)
Linnaeus filius. *Supplementum plantarum systematis vegetabilium* (1781)
J. G. A. Forster. *Characteres generum plantarum* (1776)
Michel Adanson. *Familles des plantes* (1763)
C. P. Thunberg. *Flora Japonica* (1784)
P. J. Bergius. *Descriptiones plantarum ex Capite Bonae Spei* (1767) [Cape of Good Hope]
Jan Burman. *Thesaurus zeylanicus* (1737) [Ceylon]
Rariorum africanarum (1738-1739)
Herbarium Amboinensis (1769) [Ambon Island in the Moluccas]
Charles Plumier. *Plantarum americanum genera* (1755-1760), edited by Jan Burman.[43]

42. Ibid. 1: 233-240.
43. Ibid. 1:254.

André Thouin's additional task was the selection of the gardeners for the two expeditions, and he chose two in whom he had complete professional confidence: Jean-Nicolas Collignon to be with La Pérouse, and Félix Delahaye to accompany Entrecasteaux.

IV
The Voyage of La Pérouse: Part I
From Brest to Monterey Bay

From the outset, the reader must take into consideration that the voyage of La Pérouse, sailing from Brest at the beginning of August, 1785, would end in shipwreck, presumably in the summer of 1788 in the western Pacific. His final port of call had been Botany Bay (Australia) from which he had sailed on 10 March 1788, never to be seen again. The original manuscript of his journal, obviously a fundamental document for the record of the voyage, was lost with him; as were the principal studies, the daily notes, and the major part of the collections made by the botanists and the gardener.

The reader may also recall that André Thouin, in his instructions for the gardener that had been delivered to La Pérouse, emphasized the widom of sending to Paris packets of seeds and herbarium specimens from any European outposts visited during the long voyage. He called it a precaution in the tacit knowledge that no extended venture into uncharted waters could be guaranteed a happy ending. It is unlikely that La Pérouse, a veteran of perilous seas, needed the reminder; and he would take every opportunity at ports of call up to Botany Bay to send back *copied* portions of his journal to France, along with letters, maps, drawings, seeds, and plants.

Catherine Gaziello, who examined the copied journal in manuscript, noted that some of the corrections on the pages had been made in La Pérouse's hand, but not all, leaving room for doubt that he had seen such corrections. It also seemed apparent that not all the material presumably shipped had reached its destination. Even so, she was impressed most favorably by the evident integrity with which baron de Milet-Mureau had edited the surviving portion of the journal, especially as he had been a military rather than a naval officer, thus implying the reliability of the published text. It is a pity that other journals, those of the scientists in particular, were not similarly shipped in installments; but, as La Pérouse once reminded his gardener, such documents were to be reserved for the king's eyes alone.[44]

Jean-François Galaup de La Pérouse was born near Albi in 1741, descendant of an eminent and prosperous Languedocian family. His parents bought the domain of La Pérouse for him, giving him seigneurial status; but the family was bourgeois. The title comte, usually attributed to him, is incorrect, so that his name should

44. Gaziello. 1984. *L'Expédition de La Pérouse*, pp. 19-20, 216-218.

more properly be rendered Galaup-La Pérouse no matter the customary use of the particle. He had entered the naval school at Brest at a very young age, emerging as a commissioned garde de la marine on 19 November 1756: a "blue" officer, that is, wearing the blue uniform of officers of common origin. The "red" officers wore the red uniforms reserved for men of aristocratic origin, and this traditional distinction had been prejudicial to the advancement of "blue" officers.

After the maréchal de Castries took control of the ministry of the marine in 1780, he abolished the distinction by establishing formal equality in advancement between nobles and commoners in an attempt to attract the talented to naval service. However enlightened the reform may have been, the traditional bias in favor of aristocrats did not totally disappear in practice as the choicest posts tended to be given to them. A more conclusive end to this caste system would only come with the Declaration of the Rights of Man and Citizen on 27 August 1789.[45] It seems probable that La Pérouse benefited from Castries' reform. His promotion had been slow despite a substantial service record, until he was promoted to the rank of captain in 1780 and given command of a frigate.

The following year, the government developed a plan to destroy the English posts on Hudson's Bay. The mission was confided to La Pérouse, and he was given command of a 74-gun warship supported by two frigates of 36 guns each. He slipped out of Santo Domingo in great secrecy in the late spring of 1782. Passing Resolution Island, he met ice and storms in Hudson Strait and could not pass into the bay until 17 July 1782. During the first weeks of August, he took three English posts on the western side of Hudson's Bay: Fort Prince of Wales on the Churchill River; Fort Williams on the Nelson River; and Fort Severn on the Severn River. In all instances, troops were put ashore under cover but in difficult terrain. They would block the fort from the land side to prevent retreat, while the ships would then appear off shore with obvious artillery superiority. The fort would then surrender when initially summoned to do so, in the rational, honorable tradition of the 18th century. There were no casualties.

The forts, to be sure, were then burned, but not before removing the stores and furs; and the governors of each post were taken prisoner. Because of the approach of cold weather and danger from Indians, La Pérouse left provisions and arms for the English troops so that they might make a successful retreat to another English post. The gesture won him a reputation as a humane and generous man. Beyond which, his bold exploits in 1782, at a time when French naval fortunes were at low ebb, gave him national distinction at home.[46]

45. Horner. 1995. *Looking for La Pérouse*, p. 34.

46. Milet-Mureau. 1797. *Voyage de La Pérouse*. 1: xxxiii-xlvii; and Dunmore. 1965. *French Explorers in the Pacific*. 1: 254-255.

His record, in short, indicated he would be well-suited to command the expedition in 1785, and his friendship with Claret de Fleurieu meant that he would not be overlooked. After the peace of 1783, however, and with 18 campaigns behind him, La Pérouse had been entitled to an easier existence. His acceptance of what could only be a most demanding and perilous assignment was given after appeals to his sense of duty and as an obligation to the king. Two frigates were assigned to him as commander-in-chief, and he was given the appropriate latitude to select naval presonnel in whom he had confidence. For *La Boussole*, the frigate on which he would establlish his own headquarters, he chose the chevalier Sutton de Clonard as captain. For *L'Astrolabe*, he named Paul-Antoine-Marie Fleuriot de Langle, who had been with La Pérouse in Hudson's Bay, as captain. His naval surgeons remain remarkably obscure considering the fame of the voyage. A Dr. Rollin would serve as chief-surgeon on *La Boussole*; while on board *L'Astrolabe* a Dr. Lavaux would be chief-surgeon assisted by Dr. Jean Guillon.

La Pérouse accepted the scientific personnel as recommended by the appropriate academicians in Paris. His plea that some consideration be given to congeniality in addition to professional competence does not seem to have been given much weight. Yet, a warning about such matters had been published by James Cook. During his second voyage, Cook had had the botanist Johann-Reinhold Forster on board, whose unsociable character had rendered the atmosphere difficult during the entire voyage. On land, or in the academy, we can find in space the necessary defense against the noxious personality; but what is one to do when confined for 3 to 4 years on a ship perhaps 120' by 30'? Such perils of sea travel have rarely been adequately delineated.

Joseph Boissieu de Lamartinière was the senior botanist on the expedition. He was on board *L'Astrolabe*, but precious little is known about him. He had been born in Saint-Marcellin (now Isère) sometime between 1750 and 1755. After achieving a master of arts in 1776, he enrolled that September as a medical student at Montpellier. His subsequent activity as a botanist remains obscure, but his correspondence with André Thouin at the time of his appointment suggests great familiarity, and we know that he was recommended for the voyage by Antoine-Laurent de Jussieu.[47] The letters suggest a healthy, confident young professional, very practical, impatient with dreamers and pessimists.

Nicolas Collignon, the gardener for the expedition, was born in Metz. Judged by Thouin to be active and intelligent, Collignon was only 24 at the time Thouin made the recommendation. He had only worked under Thouin's direction. The

47. Henri Cordier. 1916. "Deux companions de Lapérouse." *Bulletin de la Section de géographie (Comité des travaux historiques et scientifiques)*. 31: 55-56.

detailed instructions Thouin provided were his way to continue to instruct and guide him. It is clear that Thouin did not want Collignon to become subordinate to the botanist, but to remain independent to carry out the duties defined in his instructions. He was authorized (to occupy his time) to assist the other naturalists in seeking and preparing their plants so long as that did not interfere with his principal mission. Consequently, when Collignon reached Brest, he asked to be assigned to a ship different from Lamartinière's. Whereas, Lamartinière, with some reason, argued that the gardener ought to be subordinate to the botanist.[48] La Pérouse resolved the dispute in Collignon's favor, assigning him to La Boussole, and the issue would remain a point of friction between commander-in-chief and botanist.

As each ship necessarily had a chaplain on board, La Pérouse selected two who could occupy free time usefully as assistants in natural history. Father Jean-André Mongez (sometimes Mongès), a canon of Sainte-Geneviève, became chaplain on *La Boussole*. Father Claude-François-Joseph Receveur, of the Minimes Order, although sometimes identified as a Franciscan, served as chaplain-naturalist on *L'Astrolabe*. While it might appear from this that La Pérouse did not expect the spiritual requirements of his crews to be burdensome, he was consistently careful in selecting naval personnel to prefer men with multiple skills as a measure of efficiency.

The painter on board *L'Astrolabe*, at the botanist's disposition for drawing plants, was Guillaume Prévost, known on the expedition as the uncle. His nephew, Jean-Louis-Robert Prévost, a very young man, was on board *La Boussole* as the illustrator. Both remain obscure figures today. Another painter, Gaspard Duché de Vancy, was engaged primarily for drawing figures and landscapes, sailing on *La Boussole*. For reasons not entirely clear, Lamartinière did not find his illustrator to his liking, perhaps because Guillaume Prévost was said to be of a melancholic disposition, and this want of congeniality was evident even before they left Brest. All told, the situation within the botanical team did not augur well for the success of that aspect of the voyage.[49]

At the beginning of August, 1785, while waiting for favorable winds to leave the port of Brest, six sailors and one soldier on board were detected as infected with venereal disease, having escaped prior examinations by the surgeons, and were immediately put ashore before departure.[50] Otherwise, the first leg of the journey to Madeira and the Canary Islands was uneventful.

48. Lamartinière to Thouin, Brest, 4 July 1785. Ibid. pp. 56-58.

49. Gaziello. 1984. *L'Expédition de La Pérouse*, pp. 138, 142-144, 149.

50. La Pérouse. 1797. *Journal.* 2: 12. The form of this citation is meant to signal La Pérouse's own narrative as distinct from the editorial material produced by Milet-Mureau.

At Santa Cruz de Tenerife, the naturalists' main objective was to climb the great volcanic peak on the island. The botanist's task, Lamartinière wrote to Castries, went beyond simply collecting large numbers of plants for arrangement in a herbarium and making a catalogue of them. He must examine the soils, note down exposures and temperatures, in order, finally, to determine by analogy which of the local productions would be suitable for propagation in France and thereby render an important service to the nation.

He had encountered, he continued, a number of plants that should do very well if cultivated in Languedoc and would be sending him the seeds for seven or eight species that he believed would benefit the economy of that province: some to improve pasturage, some to provide new woodlands where there is a great shortage. In particular, he cited *Spartium supranulium* Masson, a large shrub that served the islanders as a forage for goats. *Asparagus declinatus* L., a charming shrub, and *Cistus villosus* L. [=*C. incana* L.]. Finally, *Euphorbia canariensis* L., used generally for fuel. He emphasized its formidable growth: one trunk can produce more than 150 branches as thick as a man's arm and 12' in length, enough wood to provide heat for one person all winter.

As *Euphorbia canariensis* grows in dry, rocky areas, Lamartinière recommended experimenting with it in those waste lands called *garrigues* around the village of Montferrier just north of Montpellier. He was inclined to believe that all these plants would do well in that soil, recommending that Antoine Gouan, under whom he had studied for the medical degree at Monpellier, be put in charge of such experiments.[51] The recommendation was appropriate as Gouan had been experimenting with the introduction of exotic plants as the best method to replace vegetation lost to excessive timbering.[52]

In a letter written the following day, Lamartinière exposed even more explicitly his practical disposition. He reported to Thouin that he had seen a good many plants, but all of them were described species. Yet, the absence of novelties seemed to distress him not at all as he expressed little interest in finding new genera and species. He emphasized that his collection of seeds could be useful to Thouin. There remains a short list in the Muséum d'Histoire Naturelle in Paris, dated May 1786, indicating that the seeds sent from Teneriffe had been planted in the Jardin du roi in Paris, as well as redistributed to Antoine Gouan and Pierre-Joseph Amoureux in Montpellier; the abbé Rozier and Claret de La

51. Lamartinière to Minister of the Marine, 29 August 1785. Milet-Mureau. 1797. *Voyage de La Pérouse*. 4: 247-250

52. See Antoine Gouan. 1796. *Herborisations des environs de Montpellier, ou Guide botanique à l'usage des élèves de l'Ecole de santé.* Montpellier: G. Izer & A. Ricord.

Tourette in Lyon; Dominique Villars of Grenoble; Philippe Picot de Lapeyrouse of Toulouse; Jacques-Martin Cels of Montrouge; and both L'Héritier de Brutelle and Malesherbes.[53]

After leaving the Canary Islands, the two frigates steered southward for Trinity Island (Ilha da Trindade), anticipating it to be a Portuguese station from where fuel and water could be obtained. On 17 October 1785, a longboat was lowered to go ashore to ascertain opportunities but experienced a difficult landing because of rough seas. Lamartinière and Receveur went along for botanizing but were refused permission to collect plants by the Portuguese authorities who exhibited great anxiety about French intentions. The woeful condition of the small Portuguese garrison made it immediately clear that the island could not provide significant supplies of wood or water, so there was nothing to do but to proceed to Saint Catherine Island (Ilha de Santa Catarine) off the southern coast of Brazil, where supplies were known to be available.

Anchoring there on 6 November 1785, they found a pleasantly fertile island with ample supplies and met a hospitable reception. Collignon added some orange and lemon trees to his collection of live plants for transfer to the islands of the South Seas. He also obtained seeds for cotton, maize, rice, and other staples suitable for tropical use.[54]

The navigational instructions given to La Pérouse directed him to search for an island that had been described in 1675 and named Ile Grande by Antoine de la Roche, a London merchant of French origin. Returning from Chile, he had discovered what we call South Georgia Island. Then, north-northeast of there, he had noted and named this pleasant, uninhabited island that had not again been sighted in over 100 years. La Pérouse spent much of December, 1785, searching the south Atlantic for Ile Grande, but in vain, becoming convinced it did not exist. The persistent good weather caused him to break off the fruitless quest and get around Cape Horn into the Pacific as quickly as possible. His initial target was the Juan Fernandez Archipelago, a known source of fresh supplies, but the pressure of growing shortages deflected him to the closer port of Concepcion on the coast of Chile.[55]

During the anchorage in Concepcion, January of 1786, Lamartinière had four drawings made of a dioecious liana, a creeping vine, of a genus he thought as yet unpublished. The drawings were sent to Paris and arrived, unfortunately,

53. Lamartinière to Thouin, 30 August 1785. Cordier. 1916. "Deux companions de Lapérouse," pp. 61, 70-71.
54. La Pérouse. 1797. *Journal.* 2: 25-26, 33, 40.
55. Ibid. 2: 42-49; Dunmore. 1965. *French Explorers in the Pacific.* 1: 261.

without a written diagnosis or description. The drawings were given to Etienne-Pierre Ventenat for analysis in the hope that a new genus could be named for Lamartinière that could be published with the record of the voyage. By the time the drawings were published in 1797, as Plates 6, 7, 8, 9, the new genus had already been published as *Lardizabala* Ruiz & Pavon in their *Florae peruvianae et chilensis prodromus*, Madrid, 1794.[56]

After a pleasant sojourn in Concepcion for repairs and resupplying, the expedition departed in March, sailing directly for Easter Island, where they anchored on 9 April 1786. The natives appeared to be friendly, but the French went ashore well-armed and in sufficient numbers to constitute a viable defensive force if threatened. La Pérouse's most immediate impression of the island was the evident excessive tree-cutting, sometime in the distant past, that had exposed the soils to intense sunlight and reduced the water supply. This caused him to suspect that most islands in the South Seas might have inadequate water supplies unless they possessed relatively inaccessible mountains where it would be infeasible to cut the timber. His long service on Mauritius, an island he thought strongly resembled Easter Island, had taught him that trees do not regenerate easily unless sheltered from ocean winds by other trees or encircled by walls. Therefore, disputing earlier reports, he concluded that the Easter Islanders had not been so devastated by their long-extinct volcanoes as by their own imprudence.

In any case, La Pérouse recognized that the island could offer virtually no resources to navigators. Unlike Cook and Forster who reached Easter Island after a long and difficult voyage, in need of supplies and afflicted with scurvy, only to find little water, wood, or food, La Pérouse was in excellent shape thanks to the recent sojourn in Chile. His ships, in fact, brought supplies for the islanders: goats, ewes, and pigs purchased in Concepcion; and Collignon was ready with seed and plants. Once an enclosure was established on land, guarded by armed soldiers, so that presents and livestock could be brought ashore, two exploring parties were formed to penetrate inland.[57]

One party was led by Fleuriot de Langle, captain of *L'Astrolabe*, which included Lamartinière, Receveur, and Collignon, went to the interior of the island. After visiting several plantations of yams and sweet potatoes, Collignon judged the soil sufficiently rich to warrant planting seeds. He endeavored to explain to natives that the seeds would produce roots and fruits good to eat. They seemingly understood and pointed out the best soils where they wanted to see the new produce grow. He planted cabbage, carrots, beets, maize, and pumpkins, subsequently

56. Milet-Mureau. 1797. *Voyage de La Pérouse*. 4: 265.
57. La Pérouse. 1797. *Journal*. 2: 79-85.

adding orange, lemon, and cotton seeds after indicating that they would produce trees.

They encountered no shrubs other than a mimosa and a paper mulberry, *Morus papyrifera* L. [=*Broussonetia papyrifera* (L.) L'Hér. ex Vent.], now recognized to have been an invasive plant possibly originating in Tahiti. They also saw considerable fields of a nightshade (*Solanum*) that the islanders seemed to cultivate on lands exhausted by yams and sweet potatoes. The more mountainous slopes were covered by grasses, but there were no signs of torrents or ravines. The French visit occurred during the rainy season, and the ground was moist despite the lack of running water. They estimated that only about one-tenth of the ground was cultivated and saw no cultivating tools, presuming that the natives simply used wooden stakes left over from the clearings. Langle estimated the population to be about 2,000, but believed it might well have been larger before deforestation. If the natives only had the industry to construct underground cisterns, he concluded, they could greatly improve their situation.[58] La Pérouse added his observation that the local agricultural produce, meaning sweet potatoes and yams, was owned communally.

Given the natives' propensity to steal anything removable, one might assume the French simply suspected that the natives had no sense of private property. But, the hydrographic engineer had noted that the agricultural fields, always rectangular in form, were never enclosed with either hedges or fences, suggesting ***communal access to food***. Whereas the plantings of paper-mulberry trees were always confined by enclosures, indicating that the principle of exclusion was understood; while the banana trees, obviously valued and carefully tended, were planted in quincunx but not enclosed. He reported no evidence of attempts to restore the tall trees that had been virtually eliminated. Despite the absence of surface streams, he assumed the presence of underground water having seen occasional shallow excavations amongst the surface rocks that contained limited amounts of bad-tasting water.[59]

The second exploring party, meanwhile, which included La Pérouse, remained closer to its base, but examined the unique monuments, their platforms, the natives' houses, and their plantations. They saw no evidence of a religious cult practiced nor of any attempt to repair the fallen monuments, although the natives obviously venerated them.

58. Langle's Report. Ibid. 2: 97-103.
59. Report by Gérault-Sébastien Bernizet. Milet-Mureau. 1797. *Voyage de La Pérouse*. 4: 34.

The livestock brought from Chile was distributed, but with little confidence that the pigs could survive the shortage of fresh water. La Pérouse believed that the goats and ewes, as they drink little and like salt, might well multiply. Langle told about one chief, to whom he had offered a pair of goats, who had taken them with one hand and grabbed Langle's handkerchief with the other. Indeed, virtually all those on shore had had their hats and pocket handkerchiefs stolen by natives, who would seize such items and run. The experience was dispiriting.[60]

After a day among the Easter Islanders, La Pérouse wrote: "No one who has read the recent reports by travelers can any longer take the Indians of the South Seas for savages. On the contrary, they have made very great progress in civilization; and I believe them to be as corrupted as can be relative to the circumstances in which they are found. This opinion is not founded upon the various thefts they committed, but upon the manner in which they were carried out.

"The most brazen scoundrels in Europe are less hypocritical than these islanders. All their affection is feigned; their countenance expresses not a single genuine sentiment. The Indian of whom you must be the most suspicious is the one to whom you have just made a present, and who appears most disposed to render you a thousand small services. Fathers violated the integrity of their 13 and 14-year-old girls by bringing them amongst us in expectation of payment. The evident aversion of these young Indian girls was proof that local mores were being violated in their regard. No Frenchman took advantage of the barbaric right that was offered. If there were moments given to nature, the desire and the consent were reciprocal, not for payment."[61]

To illustrate further the islanders' relative sophistication, La Pérouse noted the considerable intelligence with which they cultivated fields. Grasses were pulled out, put into piles and burned, the ashes used to fertilize fields. Their interest in the technical features of the French frigates was obvious from the care they took to measure a ship, examining its cables and anchors, the compass and the steering-wheel. Such people were far from stupid beings, quite capable of reflection and appreciating the arts of Europe. And they did not miss the fact that the French did not use their weapons against them.

"We came to their island only to do good," La Pérouse concluded. "We showered them with presents; we sowed all sorts of useful seeds in their fields. We left them pigs, goats, and ewes that should multiply, asking nothing in exchange. Even so, they threw stones at us and stole everything from us they could carry off. I decided to leave during the night and hoped that at daybreak, when they

60. La Pérouse. 1797. *Journal.* 2: 85-92.
61. Ibid. 2: 93-94.

would see that our ships had departed, they might attribute our early departure to justifiable displeasure over our treatment, and that this thought might serve to render them better.[62]

Captain Monneron, the chief-engineer on *La Boussole*, reiterated the French disenchantment while predicting what the future interest in Easter Island would be: "This island, given its position beyond the main routes of navigation, given its absolute want of water and wood, and given the condition of its inhabitants who are most graciously prepared to receive, but who have nothing to give in return, this island, I repeat, offers a vast realm for the speculations of physicists and moralists; but it cannot in any way be of interest to the various maritime powers of Europe. (12 April 1786)."[63]

The surgeon, Rollin, confessed that the brief sojourn on Easter Island had been hardly adequate to treat all the questions posed by the Société de médecine. But, at the outset, he could assert that the islanders were neither the giants reported by Jacob Roggeveen [1723], nor the thin, undernourished people suffering from scarcity as depicted by a more recent traveler, presumably Cook in 1774. Instead, he saw a rather numerous tribe, better endowed with graces and beauty than he would see elsewhere during the voyage: a people well-proportioned and attractive, the men averaging about 5' 4" in height, with little bodily hair, mostly beardless, but with armpits and sexual parts well-covered with hair; well-tanned skin, the hair generally black but occasionally blond; and they were circumcised.

He found the women well-formed, too, with pleasant, oval faces and seemingly gentle. Even the men exhibited a benevolent air with little tendency to jealousy. He could not distinguish a chief among them, suggesting a state of complete anarchy. Both men and women were nearly nude, wearing only a loincloth, though an occasional individual wore a cloth cape over the shoulders down to mid-thigh. "I do not know," he added, "whether they have any concept of property, but their conduct demonstrated little respect for that of strangers. They had such a love for our hats that in short order they succeeded in depriving us of them, making us subject to their laughter, behaving rather like naughty schoolboys."

He did acknowledge their skills, finding their houses, for their kind, well-made. They could make a cloth from the paper-mulberry, a tree in short supply on the island but evidently cultivated. Their food was limited to sweet potatoes, yams, bananas, sugarcane, a seaweed they gathered at the water's edge, and some fish, although fish were not abundant in local waters. A few chickens were evident,

62. Ibid. 2: 95-96.
63. Milet-Mureau. 1797. *Voyage de La Pérouse*. 4: 101.

their only domesticated animal; and, except for rats, there were no wild quadrupeds on the island.

The great statues were numerous around the circumference of the island. The mysterious caves at the foot of the statues were found, as recorded in Cook's report. Rollin stated that native families used the caves for the burial of their dead, and that the French party had visited them without the islanders raising the least objection.[64]

The French frigates raised anchors on the evening of 10 April 1786 and set a course for the Hawaiian archipelago. Maui came into view on 29 May, its luxurious vegetation a welcome sight. Nearly 500 pirogues came out from shore loaded with fruits and pigs, which the natives hoped to exchange for iron objects. They gave every appearance of being peaceful and gentle, carrying on trade with great fidelity. They were clearly accustomed to trading and had a fair idea of values from prior experience with Spanish and English traders. As a devoted reader of Captain Cook, and knowing his fate on the island of Hawaii, La Pérouse knew the necessity of prudence. He would remain only long enough to take on necessary supplies, 48 hours. That gave Collignon very little time to plant seeds and no time to watch their development.[65]

Both La Pérouse and Dr. Rollin were aware that in Europe the introduction of syphilis into the Hawaiian Islands had been attributed to Cook's visit and widely condemned. Consequently, they anticipated meeting native resentment against Europeans but found none when they went on shore on 30 May 1786. That alerted them to the possibility that the venereal disease was not of recent introduction.

While Maui clearly provided the inhabitants with an abundance of nourishing staples, they seemed far from enjoying as good health as the inhabitants of Easter Island where the staples for existence were less abundant; and they were clearly less endowed with grace and beauty than the Easter Islanders. The men, if somewhat more robustly built, their muscles more strongly developed, were also more afflicted by diseases. Their beards were bushier, they were hairier on their bodies and sexual parts, and were not circumcised. Yet, some of them practiced a kind of infibulation by retracting the foreskin behind the glans of the penis and fastening it there with a binding. They averaged about 5' 3" in height and had little fat.

64. Ibid. 4: 7-11.
65. La Pérouse. 1797. *Journal.* 2: 111-119.

The women, too, were compared unfavorably to those on Easter Island, having neither their high spirits nor elegant bodily form, but unrefined features, a subdued air, and rather heavy and awkward in manner. Nevertheless, the inhabitants of Maui seemed pleasant, obliging, and exhibited a kind of politeness to strangers. And the beauty of the scene would have been complete had it not been for clear evidence of widespread, serious diseases: venereal disease and leprosy in particular.

Common symptoms he observed: Buboes (inflammatory swelling of lymph glands in the groin); disfiguring scars produced by suppuration; running sores with bone decay; protruberances, exostoses (bony outgrowths from bones); lacrymal and salivary fistulas and tumors; scrofulous engorgements; profound ophthamias; ichorous ulerations of the conjunctivas; atrophy of the eyes; blindness; scaly skin, pruriginous and encrusted; and a painless swelling of the extremities. Among the children, he noted head scabs or a malignant tinea that oozed in a smelly and corrosive liquid. "Most of these unfortunate victims of lust," Rollin commented, "were weak, indolent, despondent, and affected by rickets by the time they reached 9 or 10 years."

As for the painless swelling of the extremities, noticeable among the islanders on Maui and observed by William Anderson throughout the South Seas Islands, Rollin called it a symptom of an advanced elephantiasis, a form of leprosy. His diagnosis was based upon prior examination of a great number of lepers in the lazarets of Madeira and Manila.

Because Dr. Rollin was convinced that both the nature and the quality of the diet were contributing factors to health and disease, he examined several of the pigs from Maui, as pork was a principal part of the local diet. He found considerable evidence of ladrexia, a pork leprosy or pork measles. If it does not render the pork totally unfit for consumption, it produces meat that is tasteless and makes a bad meat soup. The skin of afflicted pigs was rough, encrusted, and totally devoid of bristles. In dissection, the fat was found to have a scattering of tubercles, but the viscera were found to be so full of them as to be revolting to the least delicate of men. We are left to wonder about the disposition of the nearly 300 pigs that had already been brought on board the ships before this cheerful necropsy.

As for venereal infections, Rollin observed that, "among the diseases whose ravages afflict these islanders in so deplorable a manner, one appears to be produced by the venereal virus [bacterium] in pure form; but most often it seems to appear in an altered character or in combination with psoriasis." As it was common in the 18th century to consider all venereal infections as varieties, or stages, of syphilis, including gonorrhea, in the absence of more detailed symptoms we cannot be certain what Rollin saw beyond syphilis in pure form. Moreover, time and circumstances did not allow him to make any inquiry about the treatments used by the natives to counter any of these ailments. "But if I were to judge solely by their surrender to the pain," he surmised, "and by the evident progress of their infirmities, I should have to believe that they know no effective means to put an end, or even to provide any easing, to so miserable a condition."

What of the charge that syphilis had been spread to the Hawaiian Islands by Captain Cook's crews? Cook had visited Maui about 9 1/2 months after having been ashore on Kauai and Nihau. He had initially meant to preserve Maui from the spread of venereal disease by not allowing his crews to go ashore, only to discover that syphilis was already present. That implied to Rollin that, if Cook's crews had been responsible for introducing syphilis to Kauai and Nihau, the disease would have had to be transmitted to Maui by the natives themselves within that 9 1/2-month period. Given what was then known about the etiology and the manifestations of syphilis, plus the malformations that he saw on individuals of all ages on Maui, one could only conjecture, if not prove, that venereal disease had existed on all those islands long before Cook's visit: either because the disease was indigenous or because it had been brought by navigators who preceded him.[66]

Dr. Rollin made no comment on the possible practice of cannibalism on the Hawaiian Islands as Cook and Anderson had believed. La Pérouse, finding the natives so amiable, could not believe that they indulged in such uncouth consumption despite clear evidence of cannibalism elsewhere in the South Seas. Instead, the landing on Maui moved him to speculate about the "foolish tradition" of laying claim to the land of others in the name of one's king. He thought the practice completely ridiculous. Because men have possessed artillery and bayonets, he added, they have treated fellow human beings with contempt: not respecting their most sacred rights; regarding the lands irrigated by the sweat of their inhabitants, lands that had served for many centuries as the burying ground for their ancestors, simply as an object of conquest.

"These people date back to those happy times when religion did not serve as a pretext for violence and greed. When describing the mores of newly-discovered peoples, the contemporary navigator should have no other goal than to complete the history of mankind. Such voyages must end in the knowledge of the globe. The enlightenment they seek to extend is meant for nothing more than to make a happier existence for the islanders they visit and to augment their means of subsistence. Accordingly, contemporary navigators have transported bulls, cows, goats, ewes, and rams to these islands; they have planted trees, sown seeds, and brought the tools to accelerate industrial progress. For us, we would be amply

66. Milet-Mureau. 1797. *Voyage de La Pérouse*. 4: 12-19.

recompensed for the extreme fatigues of this campaign if we could succeed in destroying the practice of human sacrifices that are said to be widespread among the islanders of the South Seas."[67]

La Pérouse adhered to his decision to make the Hawaiian visit brief, and the French were at sea again by the evening of 1 June 1786, heading for the Alaskan coast. They got their first view of Alaska on 23 June when they recognized Mount St. Elias, snow-covered, in the background of a coastal chain of mountains. The most immediate good anchorage would have been Yakutat Bay, but they turned southeast in search of a more sheltered and wooded site. On 2 July 1786, they entered a bay that seemed more suitable, perhaps because it had a small island in it that could serve as a base for an observatory. La Pérouse named it Port des Français, later Frenchmen's Bay, and now Lituya Bay.[68]

The ships were soon surrounded by canoes, the natives appearing to be experienced traders, offering fish and furs in exchange for iron, a metal not unknown to them. This suggested probable earlier visits by either Russian or American travelers, or even a possible overland connection with Hudson's Bay by following a river that emptied into the bay from the east. It became immediately evident that otter furs were unusually abundant in the region. Aware that the Indian villages were all on the mainland, the French established their base on the island. Their sense of security was short-lived. The Indians proved to be accomplished thieves, and the post had to be abandoned. That did not prevent the gathering of considerable wood and drawing a good supply of water.

On 13 July 1786, three small boats, being employed for sounding and charting the interior of the bay, were caught in an unexpected squall. Two of them were swamped in a ferocious turbulence, resulting in the loss of 6 officers and 15 crewmen. The catastrophe was a devastating blow to an expedition that, after nearly a year at sea, had virtually not even had an illness on board. They would not sail on until 30 July 1786, when all hope of finding survivors on shore was extinguished.

During the interim, Lamartinière was able to make a number of field trips to study the vegetation. Even though the region was on the latitude of Hudson's Bay, the climate was much milder. Plants enjoyed vigorous growth for 3 to 4 months a year, making it evident that Russian wheats and many other useful plants could flourish in the region. Such an abundance of edible wild plants was found that they were brought on board to be put into soups, stews, and salads, contribut-

67. La Pérouse. 1797. *Journal*. 2: 124-125.
68. Dunmore. 1965. *French explorers in the Pacific*. 1: 266-267.

ing to maintaining good health: a wild celery, probably *Ligusticum scoticum* L.; a lupine, probably *Lupinus nootkatensis* Donn; milfoil, *Achillea borealis* Bong.; a wild chicory that remains obscure; and a mimulus, likely *Mimulus guttatus* DC. Many other common plants had clear generic affinity with those in the meadows and mountains of France: angelica, buttercup, violet, and several grasses appropriate for forage; as well as other herbaceous plants that could have been collected and eaten without danger had they not been enmeshed with the roots of a very hardy hemlock unknown to him, but obviously to be avoided, probably *Cicuta douglasii* (DC.) Coult. & Rose. The woods were filled with strawberries, raspberries, and gooseberries. He found elderberries, various species within the Heath Family (Ericaceae) growing in the shade, dwarf willows, balsam poplar, sallow willow, hornbeam-birch, and superb pines that could provide masts for the largest ships in the navy. Because of the great similarity between the plants of that region and those of Europe, Lamartinière believed he had encountered only three species entirely new to him.

La Pérouse believed that the natives' ignorance of the nutritional value of the plants around them was a valid measure of their primitivity. The streams were full of trout and salmon, their primary food. The natives, who were probably Tlingit Indians, sold the French the skins of black and brown bear, Canadian lynx, ermine, marten, grey squirrel, beaver, Canadian marmot, and red fox, surely indicating an available supply of meat. The most common, yet the most valued, of the furs were those from sea otter, wolf, and seal.

In only one respect did La Pérouse believe those Indians had achieved any civilizational progress: he described their arts as quite advanced. But he saw little or no progress in the refinement of mores or in the containment of ferocity. He could detect no subordination of any individual to any higher authority. Every Indian seemed to be agitated continually by fear or by a thirst for vengeance. To be sure, he saw them only in summertime, recognizing that during the harshness of winter, when hunting would be difficult, they would be continually exposed to starvation. In the summer, when there was great abundance, when it required less than an hour to provide a family's supply of fish, the men were free to spend the remainder of the day gambling: a pastime for which they exhibited a mad passion, and which was the source of continual quarreling. La Pérouse predicted that, if ever such a people should become acquainted with any alcoholic beverages, it could destroy such a people entirely.

"The philosophes," he added, "who write their books in comfort, depicting such peoples as good, because they are close to Nature, have not seen the injustices and the deceit that I have seen during 30 years of travel. Although confirmed in this opinion by sad experience, I have nevertheless believed myself obligated to employ the powers whose direction has been confided to me to remove the injustices from these primitive lives and to make them understand that justice is a right of mankind that can never be violated with impunity."

He could find no morality in that Alaskan people. No matter how many kindnesses were shown them, or presents given to them, they showed themselves completely insensible to benevolence and seized every opportunity to steal from the French. Quarreling continually amongst themselves, indifferent to their children, the men were also true tyrants with their women, who were condemned to unceasing and difficult labor. While La Pérouse readily admitted that it should be impossible to conceive of a society without some virtues, he confessed his inability to find any in that one. Nothing softened the picture. We are given a vivid description of Indians' filth, of cooking pots never washed, and dwellings unendurably stinking. If they were perhaps more advanced in their arts than the inhabitants of the South Seas with the exception of agriculture, it occurred to La Pérouse that perhaps agriculture was the more critical advance in softening the mores and rendering a people more sociable.[69]

When the expedition sailed southward from Lituya Bay at the end of July, no one on board could recommend the site as a desirable port of call or as a suitable place to establish a trading post in possible competition with Russian and Spanish traders. Given its distant location from France, and the probable high cost of providing adequate security, any profits to be derived would not be commensurate. That conclusion was so obvious, Captain Monneron avowed, that any lengthy report on the subject would be a waste of time.[70]

From the botanists' viewpoint, the delayed departure from Lituya Bay was the more regrettable, because the expedition could only reach Monterey Bay, the next destination, in mid-September, by which time many of the local plants had released their seeds. They were struck at once by the evident great fertility of the soil. Given the timing, it made sense to donate seeds brought from Paris to the Spanish governor and to the mission for their gardens, and the seeds were gratefully received. Collignon also gave the missionaries some potatoes he had obtained in Chile, still in a perfect state of preservation, confident they would do well in the region around Monterey. In a region where the harvests were very bountiful compared to those of Europe or Chile, the French were surprised by the rarity of fruit trees in such a suitable climate. The governor responded to the gifts very generously with supplies of cattle, vegetables, milk, and chickens for the ships, and

69. La Pérouse. 1797. *Journal.* 2: 136-205.
70. Milet-Mureau. 1797. *Voyage de La Pérouse.* 4: 102-103.

assisted in obtaining wood and water. The French even had to insist on paying for supplies.

Behind this facade of amicability, the French profoundly disapproved Spanish treatment of the local Indians and made a substantial attempt to understand the Spanish regime and its rigorous control of Indian life. "The missionaries remain persuaded," La Pérouse noted, "either because of their prejudices or perhaps because of their own experiences, that reason is rarely ever developed in these people, which becomes for the missionaries a justifiable excuse for treating them like children and admitting only a very few of them to communion. Such is the disposition of this small band of religious who, like Descartes and Newton, could have enlightened their century and their charges by teaching them that four plus four equals eight, a calculation now beyond the grasp of most of them. The regime within the missions here is not adapted to bringing them out of that state of ignorance. Everything is devised to recompense the Indians in the next life. The most common of arts, such as the surgery available in French villages, is not available to them. Such a conception of a community, originally devised by Jesuits but imitated by succeeding religious orders in California, is contrary to the progress of civilization, no matter the piety and charity of these religious."[71]

Robert de Paul, chevalier de Lamanon, a physician-naturalist on board *La Boussole*, who had endeavored to learn some basic vocabulary of the Indians of Lituya Bay but without much success, worked on several Indian languages in the Monterey area. Again, he found the local languages very difficult; but, in this case, he drew limited help from several Spanish-speaking Indians, enabling him to recognize that the local Indians, if living close together, actually represented more than one tribe, each preserving a separate idiom. Lamanon also noticed that the Spanish missionaries were entirely dependent upon native interpreters for local communication, revealing a mentality that a man of the French Enlightenment could only deplore.

In the short time he had within the Mission of San Carlos, founded by the Franciscans in 1770, Lamanon distinguished two linguistic groups. The language of the group he called Achastliens reflected limited intellectual development, that is, few abstract conceptions. They made, for instance, little distinction between the genera of animals, using the same word, *ouakeche*, for frogs and toads. They did not distinguish by name the plants they used. Their epithets for moral qualities were nearly all borrowed from the sensations of taste. The word *missich* indicated both a *good* man and a *savory* food. Their word *keches* meant both a *wicked* man

71. La Pérouse. 1797. *Journal.* 2: 274-275.

and *rotting* meat. They used their fingers to count up to 10, seemingly unable to count from memory or independently. Yet, they had words for 1 through 10. The nearby group he called the Ecclemachs had an entirely different language, much more sophisticated grammatically, a much richer idiom, more akin to a European language. As a result, Lamanon was able to transcribe a somewhat more extensive vocabulary for them.[72]

The chief-surgeon, Rollin, had kept notes on his observation of the American Indian population: very limited in Chile, somewhat more extensive in Alaska, but chiefly at Monterey Bay, where he examined a great number of Indians of both sexes. Certain European authors, he asserted, but without mentioning any names, have described the American Indians as a degenerated species, even suggesting the inevitable degradation of Europeans naturalized in America. Simple flights of imagination in his opinion, and no more on target than their view of the supposed degeneration of animals transplanted from the old continent to America (the opinion of Buffon). As for similar assertions meant to illustrate such degeneration, he said, "I have nowhere seen the prolongations of the scrotum, nor the prodigious swelling of the penis, nor men whose nipples provide milk as has been reported by some travelers: no evidence that uncivilized people can run faster or have sharper sensory organs than Europeans."

On the contrary, he asserted, among such peoples the course of life is much like ours. The periods of growth and decline are the same everywhere; and perceived differences, he argued, merely reflected the effects of climate and the conditions of life. While Rollin readily recognized tribal differences in physical appearance, attractiveness, and the marked differences in the distribution and quantity of bodily hair, he was equally struck by living habits that most tribes had in common: as important in characterizing Indian life as the differences, which were perhaps more superficial.

Certainly he regarded the Alaskan Indians visited in Lituya Bay as unique in their thievery, their brazenness, their excessive bad temper. Living on fish and game, with an easy abundance available during the summer, they actually fed on spoiling or nearly rotten food in preference to taking the modest trouble to obtain better food. Their laziness was also revealed in their crude food preparation. Either they simply grilled directly on coals, or they boiled in wooden mess-bowls filled with water into which they tossed stones, reddened in the fire, until the cooking was completed.

72. Ibid. 2: 288-291.

While that degree of crudity was not evident in the natives of the region around Monterey Bay, Rollin remarked on Alaskan and Californian similarities. Neither group made any significant use of vegetable foods beyond a few pine nuts or fruits in season, never forming an essential part of their diets. Both groups tended to sleep pell-mell on skins or mats thrown around a fire without distinction of age or sex. He saw indication that such peoples were polygamous; that is, that marriages lasted only so long as they were convenient to both parties. They seemed to attach little importance to the exclusive possession of their wives, often seeking some advantage from their favors, trading them for a few pieces of iron or some glass beads.

Rollin acknowledged that his ignorance of the languages around Monterey Bay had greatly increased the difficulty of describing local diseases satisfactorily. But the length of the anchorage there gave him greater opportunity to observe a considerable number of Indians, and he had benefited from the assistance of the chief-surgeon assigned to the colony, as well as from information provided by a missionary. The summary of his observations suggests a conscientious effort to minimize cultural barriers and a compassion for the natives' plight:

Common wintertime maladies: sore throats, catarrhous complaints, pleurisies, and pleuro-pneumonias. Remedies: herbal teas usually, after which the herbs or plants were ground and placed on the epiglottis or the site of the pain. If these maladies reached a certain degree of seriousness, they commonly degenerated into chronic diseases because of the inadequacy of the remedies. Their terminal illnesses would likely be either pneumonia or consumption, that is, emaciation. Spring and autumn maladies: Transient, intermittent fevers and accumulations of mucus were particularly evident. He could not determine whether the local tribes knew of any substitute for quinine to combat fevers. Their general practice seemed to be limited to forcing a finger down the throat to provoke vomiting and using steam baths to induce abundant sweat.

Summer diseases: The putrid fevers, a generic name given to fevers attributed to the breakdown of the humors, because the breath and the excretions of the sick person gave off foul odors; typhus; burning fevers; bilious fevers; and dysentery. The usual inability to cope with any of these fevers made them especially distressing. For if the efforts of Nature were insufficient to provoke salutary evacuations, whether by faeces, urination, or perspiration, the patients usually became victims. The most fearful of such fevers were the burning and the bilious. Their development was so violent that stricken individuals rarely had the strength to resist.

Beyond such seasonal maladies, Rollin noted what he called nervous fevers, rheumatisms, psoriatic ailments, ophthamalias, syphilis, and epilepsy. He was told that there had been epidemics of measles and smallpox in the past, introduced by several European ships, and murderous for the native population. While it was understood in Monterey that syphilis had initially been taken from America to Europe by Columbus's fleet, reliable people in Monterey believed that the disease had been unknown around Monterey until Europeans reached the area.

The natives had then developed two specific treatments for syphilis: a sand bath they called the *tomascal,* and a decoction derived from a plant and taken as a drink. For the bath, a ditch would be dug in the sand, to be heated by a fire built within it. After the fire had been removed, the patient would lie down within the hot ditch and be covered with sand up to his neck. The intention was to induce sweating. When the sand bath had cooled, the patient could be removed and washed off in the sea or a nearby stream. The bath would be repeated until a complete cure should be achieved, meaning presumably until all the primary and secondary symptoms of syphilis had vanished.

The plant commonly used to treat venereal disease was known by the local Spanish as *gouvernate,* a resinous shrub to which Rollin did not give a botanical name. He did provide a description, probably with the assistance of Lamartinière:

> Calyx: 4 ovoid segments, similar to the petals, inserted under the ovary.
> Corolla: Polypetalous; 4 petals, small, entire, oval, inserted on the receptacle.
> Stamens: 8, inserted on the receptacle; filaments fleshy; anthers simple.
> Pistil: Ovary enclosed, pubescent, pentagonal, divided into 5 carpels, each enclosing 1 oblong seed; the hairs on the pericarp very evident but quite fine.
> Habit: Stalks angular, clustered, gnarled, coated with a sticky glaze; lateral branches alternate and close to one another; leaves small, petiolate, bilobed, opposite, smooth above, the nerves hardly evident beneath; flowers axillary or terminal, pedunculate, solitary or occasionally in pairs.

His description strongly suggest his dried specimen was the occasional 4-merous, if usually 5-merous, *Larrea tridentata* (Sessé & Moç. ex DC.) Coville, the creosote bush, a dominant shrub widespread in dry areas of California, whose resinous odor is a characteristic. The species would not be published until 1824, and then as a *Zygophyllum.*

Rollin also took an interest in childbirth and abortions, and subsequent ailments such as uterine bleeding and other losses of blood. He was impressed that the Indian women experienced very little inconvenience during pregnancy, and most gave birth with ease. Laborious accouchements were so rare as to be regarded as unnatural; but when they did occur, both mother and child usually perished. In natural childbirth, the expulsion of the infant was preceded by pains only briefly. Once the birth was accomplished, older women assumed the duties of midwives: tying up the umbilical cord and washing the infant in cold water. Suckling time seemed to vary considerably, sometimes for a very short period, but commonly from 18 to 20 months.

The concluding remarks in Rollin's report could not conceal, despite his sympathy, a picture of Indians virtually bereft of effective medical treatments. While he had detected some specific remedies for particular afflictions (hot sand-baths, cold baths, and certain plants) recommended by medicine men (whom he characterized as charlatans who had gained general confidence), their most common remedy was to arouse sweats as the purification favored by Nature.

He had observed that hernias were very common. The Indians had apparently no method for repositioning the tumors nor used any bandaging to keep them in place. Rollin performed such procedures on several children in the presence of fathers and mothers in the hope of teaching the procedure, as well as explaining how such accidents could be avoided. Their evident lack of comprehension left him in considerable doubt about the efficacy of his efforts. If the Indians seemed to have little idea about what to do for dislocations, they were somewhat more rational in coping with fractures. Broken bones were reset, held in place with bandages and covered with tree bark, the patient kept immobile during the healing.[73]

At least some of the plants and seeds collected in Alaska and California would be sent back to Thouin from the next port of call, Macao, by Collignon and Lamartinière. The visit to Monterey, therefore, marked the beginning of the botanical exploration of the Trans-Mississippi West. The true extent of Collignon's collection remains obscure, partly because he collected some seeds from plants too late in the season to identify; partly because he would not have recognized some distinct American plants, giving La Pérouse a short list of them using common names, some of which were likely his own invention. The following is a list of probable identifications:

1. L'absinthe maritime (coastal sagewort) = *Artemisia pycnosephala* DC.
2. Le grand absinthe (California sagebrush) = *Artemisia californica* Less.
3. L'armoise (common sagebrush or mugwort) = *Artemisia douglasiana* Besser
4. L'aurone male (silver wormwood) = *Artemisia ludoviciana* Nutt.

73. Dr. Rollin. 1797. "Mémoire physiologique et pathologique sur les Américains." Milet-Mureau. *Voyage de La Pérouse.* 4: 36-60.

5. La verge d'or du Canada (more likely California goldenrod) = *Solidago californica* Nutt.
6. L'aster (oeil de Christ) (probably) = *Aster chilensis* Nees.
7. La morelle à fruit noir (nightshade) = *Solanum douglasii* Dunal
8. La perce-pierre or criste-marine (a Medierranean umbel, *Crithmum maritimun* L.) possibly = hemlock parsley, *Conioselinum pacificum* (S. Wats.) Coult. & Rose
9. La menthe aquatique (mint), likely a variety of *Mentha arvensis* L.
10. Le thé du Mexique (Mexican tea) = *Chenopodium ambrosioides* L.
11. La millefeuille (milfoil) = *Achillea millefolium* L.[74]

Among the seeds sent back to Paris, some produced a new plant in the Nyctaginaceae (the Four-O'Clock Family), which Antoine-Laurent de Jussieu recognized as a new genus. He named it *Abronia* and would publish it in his *Genera Plantarum* in 1789. The type species would be published by Lamarck in 1791 as *Abronia umbellata* Lam., the sand-verbena, presumably from the region around Monterey: the first species from the Trans-Mississippi West.[75]

Lamartinière listed four prominent trees in the Monterey area using either established European or eastern North American names. Thus, he missed the opportunity to name, as new, what appears to have been *Aesculus californica* (Spach) Nutt., the California buckeye; *Ceanothus griseus* (Trel.) McMinn, the Carmel ceanothus; *Quercus agrifolia* Née, the coast live oak; and *Pinus radiata* D. Don, the Monterey pine. But he had, indeed, told André Thouin more than a year earlier that he had a practical disposition, an interest in useful plants, and was not primarily concerned for novelties. Yet, he knew that the two tin boxes of seeds to be sent from Macao would please Thouin.[76]

74. La Pérouse. 1791. *Journal* 2: 282. Also see S. B. Dakin. 1954. *The Perennial Adventure, a Tribute to Alice Eastwood, 1859-1953*. San franciso: California Academy of Sciences, pp. 45-46, where plant identifications differ in several instances.
75. Lamarck. 1791. *Tableau encyclopédique et méthodique des trois règnes de la nature. Botanique*. Paris: Chez Panckoucke. 1: 469; Milet-Mureau. 1797. *Voyage de La Pérouse*. 4: 251; Arthur Cronquist et al. 1972. *Intermountain Flora*. New York & London: Hafner Publishing Co. 1: 41.
76. Lamartinière to Thouin, 7 January 1787. Cordier. 1916. "Deux compagnons de Lapérouse," pp. 80-81.

V
The Voyage of La Pérouse: Part 2 From Monterey Bay to Botany Bay

The expedition sailed from Monterey Bay on 24 September 1786 meaning to cross the Pacific to reach the Portuguese colony of Macao. By that date, the Spanish enthusiasm for discovery had flagged, and their chief concern had become the security of the route from Acapulco to Manila. Whereas the French, primarily interested in new discoveries and the promotion of navigation in little-known areas, chose a more northerly route through waters less explored. The benefits from a leisurely passage for exploration had to compete with a need for repairs to the ships that could not be accomplished until reaching Macao. As a consequence, anchorages during the crossing would be few and brief, a circumstance necessarily irritating to the naturalists on board but beyond their control.

Even so, the chevalier de Lamanon was already preparing a report to the Navy, which he expected to send from Macao, which expressed considerable satisfaction with his substantial collection of fish, shells, insects, and descriptions of animals already made. But even more, he noted: "What will always distinguish this voyage, and will redound to the glory of the French nation in the eyes of the philosophes, our contemporaries, and posterity will have been our visits to peoples reputed to be barbarian without having spilled one drop of blood. The campaign, of course, is not completed; but our commander's views are known to me, and I have seen how he is supported. . . . To the qualities of skillful navigator and warrior, another merit belongs to M. de La Pérouse, one that is even dearer to his heart: that of being, at the extremities of the world, the worthy representative of humanity and of the virtues of his nation. Our voyage will prove to the universe that the Frenchman is good and that man's nature is not bad."[77]

After leaving Monterey Bay, they did not sight land again until 5 November 1786, a small, uncharted, treeless rock covered with grass. La Pérouse named it Necker Island, a point in the western part of the Hawaiian archipelago. The sea was much too rough to attempt a landing. From soundings, they recognized that the exposed rock was only the tip of a much larger submerged mass, and the area is today called the French Frigate Shoals. Pushing on westward into the Mariana Islands, they anchored off Asuncion, a volcanic island. A small party risked the difficult landing to get on shore, enabling Lamartinière to make a substantial col-

77. Robert de Paul de Lamanon to Minister of the Marine, 1 January 1787. Milet-Mureau. 1797. *Voyage de La Pérouse.* 4: 163-164.

lection of plants. In particular, he returned with several species of banana plants previously unknown to him.

La Pérouse, quite aware that seasonal changes in prevailing winds would soon trigger the departure of ships bound for Europe and waiting in Macao, and anxious to send reports of the expedition to Paris, spared no more time in his haste to approach the Chinese coast. The frigates reached Macao on 3 January 1787 and took Chinese pilots on board. They met a gracious reception from the Portuguese governor for an extended period of refitting and resupplying. But one can imagine the dismay when it was discovered that no European ships had brought them letters or packages from home. Apparently they had been sent on a ship that never arrived.[78]

Perhaps that disappointment fed the fires of discontent that flared briefly during the sojourn in Macao. Lamartinière was preoccupied writing lengthy letters to Thouin and the Maréchal de Castries during the first week, essentially progress reports on collections made since leaving Brest 18 months earlier. He estimated that they had been at sea for 15 months, and at anchor for only 3 months, to the detriment of the work of botanists and mineralogists on board. La Pérouse was quite aware of that situation and hoped to do better during the second portion of the voyage.

Lamartinière had reported from Teneriffe that he had found little that was not already known. Only when they reached Saint Catherine Island off Brazil did they meet a terrain still open to research in natural history. But the sojourn there, he now reported, had been too short, and the rainfall too heavy, to enable him to make a sizeable collection. At Concepcion in Chile, he had seemed obsessed by the opportunity to correct errors in a medicinal history of plants by Father Louis Feuillée, *Histoire des plantes médicinales qui sont le plus en usage aux royaumes de Pérou et du Chili* (1714-1725). That work contained about 100 well-illustrated plants, several of them new including the genus *Fuchsia* L.; and Lamartinière praised the work despite the errors he found. He neglected to mention in his letter the drawings he had sent back from Concepcion.[79]

Lamartinière had had only one day on Easter Island, botanically disappointing as a volcanic island largely covered by one grass, paspalum, and featuring one mimosa that was becoming rare because of constant cutting to meet the islanders' daily needs. "I do not think them capable of thinking about the future," he noted. "They are occupied with the present moment. Those we saw could not have

78. La Pérouse. 1797. *Journal* 2: 293-317

79. Report by Lamartinière, 9 January 1787. Milet-Mureau. 1797. *Voyage de La Pérouse.* 4: 172-175.

been in higher spirits, nor could they have fewer resources." He only saw there the remnants of a great revolution and felt much sadness for the poor surviving population.

One of the few plants he noted from Maui he called a *Cordia*, which the natives protected with great care, not allowing the visiters to gather any flowers from it, which suggested either medicinal or magical virtues. He was more impressed by the natives' outrigger-pirogues, "of the greatest elegance and of marvellous invention." Their cloth, made from the paper mulberry, provided a great variety of colors and designs. He lamented that the brief stopover on Maui had prevented him from learning about the different substances from which the natives drew their colors. For all botanical purposes, the visit to Maui had been unproductive.

The anchorage off southern Alaska, if lengthier, was hardly more productive. Quantities of sorrel, *Oxyria digyna* (L.)Hill, had been gathered for the benefit of a crew experiencing some scurvy. The yellow monkey-flower, *Mimulus guttatus* DC., abundant along the streams flowing down from the mountains, had been employed following the recommendation of Father Feuillée, who had claimed that Indians ate it in their soups. It would appear, in sum, that the two boxes of seeds from Monterey amounted to a confession of meager accomplishment after 18 months of roughriding at sea. "I am not very rich in plants," Lamartinière admitted, "but I have hope of doing far better when we reach Nouvelle Holland (Australia), a country absolutely new."[80]

It had been understood from the outset of the expedition that its priority would be scientific discoveries; but the instructions for La Pérouse, written by Fleurieu, quite openly expressed emphasis on the nautical sciences: a priority over the physical, natural, and human sciences. Those instructions had to weigh on La Pérouse's decisions. The length of a sojourn in any port of call generally depended more upon the requirements of the hydrographical engineers and astronomers than upon the desires of botanists, zoologists, and mineralogists. What is more, the time at sea was a dead loss for the latter group, less so for the former. The opportunity was certainly there for minor irritations and jealousies to loom larger than might have been the case in circumstances less confined than those on a small ship over many months.

La Pérouse had been aware of such tensions from the beginning of the voyage and knew, from accounts of earlier voyages, that they were to be expected. They broke into the open only once, during the anchorage at Macao, when Lamanon,

80 Lamartinière to Thouin, 7 January 1787. Cordier. 1916. "Deux compagnons de Lapérouse," pp. 71-81.

Lamartinière, Mongez, and Receveur addressed a collective letter to La Pérouse complaining about the little consideration given them. La Pérouse responded firmly, putting the four under arrest and confining them to quarters, but for only 24 hours. He did not want, as he explained in a letter to Fleurieu (24 January 1787), for them to allege that he had prevented them from making their scientific observations. He would give no further testimony of trouble.

To be precise, La Pérouse did not complain to Fleurieu about individuals in particular, but rather about philosophes, savants, and men-of-letters in general, a class a later century would call the *intelligentsia*, as inclined to dogmatic assertions. He did not aim at the scientific specialist whose competence in his own field was unquestionable, but at those he called *amateurs éclairés* who based their ideas upon philosophical presuppositions, giving them authority to pronounce on matters about which they were entirely ignorant. And, if he did not complain about specific scientists on his staff, he was sensitive to their readiness in general to make purchases of equipment recklessly since the money was coming from the king; to their routine criticisms of all religious manifestations; and to their inclination to praise the glory of the aborigines in contrast to the misfortunes of the civilized. He would have more to say on that topic in the months to come.[81]

Sufficient repairs had been completed by 5 February 1787 to permit sailing on to Manila, a comparatively short passage across the South China Sea, where additional repairs and re-supplying could be completed. They anchored in Cavite on the south shore of Manila Bay on 28 February 1787. Communications between Manila and the Macao-Canton area of southern China were frequent, enabling the expedition to receive news virtually every week. La Pérouse was surprised to learn that, soon after his departure from Macao, two French naval ships had arrived in the river at Canton under the command of Entrecasteaux. Once communication was established between them, Entrecasteaux was able to detach one officer and four crewmen for each of La Pérouse's ships, helping to fill the void created by the disaster in Alaska. The news they brought from Europe unfortunately was nearly a year old, dated 24 April 1786.

The French reaction to Spanish administration of the Philippine Islands illustrated plainly the mentality of the Enlightenment. La Pérouse found the native Filipinos to appear inferior in no way to Europeans: intelligent farmers and craftsmen, friendly and hospitable. "Although the Spanish speak of them and treat them with contempt, it was obvious to me that the vices they impute to the Indians must be attributed to the government that has been imposed upon them."

81. Gaziello. 1984. *L'Expédition de La Pérouse*, pp. 197-198.

While the initial Spanish establishments in the Philippines in the 16th century could be attributed to the greed for gold and the quest for spices, when those avaricious motives had been disappointed, they had been superceded by a zeal to spread religion. "If this zeal to preach Christianity had been enlightened with a bit of philosophy, it no doubt would have been the most appropriate system to consolidate the Spanish conquest. . . . But they only think of making Christians and never about creating citizens." Having remarked about the rigorous and brutal religious discipline (he had witnessed the penitents in the streets during Holy Week), the colonial regime's tyrannical control of the local economy, and the political despotism, he concluded that this most beautiful and most delightful country in the universe would certainly be the last place a free man would choose to live.

His own departure from Manila was delayed by the suspension of all activity during Holy Week, but the expedition finally put out to sea on 9 April 1787.[82] They headed for the Siberian coast so as to reach, in the summer, a region not previously visited by Captain Cook. On 23 June 1787, La Pérouse selected a bay that offered a protected anchorage, which he named the Baie de Ternai, known today as Ternej, in that southern portion of the Maritime Province known as Sihote-Alin. Parties went ashore at once to obtain fresh water and to seek edible herbs. They gathered large quantities of small onions, some plants in the celery or umbel group, and the familiar sorel, *Oxyria digyna*. The flowers in bloom, especially wild roses and lilies, were reminiscent of France. The botanists, in fact, found a remarkable similarity between the local plants and those of Europe, albeit only a generic community to be sure.

The great depth of the grasses not only impeded their movement on shore, but concealed whatever wild game that may have been present. Consequently, the crewmen began to work the coves of the bay with seines to obtain fresh food: cod, grunts, trout, salmon, herring, and flounder. One of the fishing parties discovered a burial site on the bank of a stream that flowed into a cove. As no native settlements had been discerned, only temporary shelters and remanants of fires, it had already been assumed that nomadic peoples had passed through the region. The grave was opened. Two bodies were found, side by side, in fairly crude wooden caskets. The heads were covered with cloth caps, the bodies enveloped in bearskin. As no decomposition had taken place, it was assumed the burial had occurred in the recent past.

82. La Pérouse. 1797. *Journal*. 2: 347-367.

Objects perceived within the grave suggested these nomads had some trade with the Chinese Empire; there were Chinese coins and copper ornaments, blue glass beads scattered in the coffins, silver ornaments that proved to be earrings, an iron axe and knife, a comb, and a small bag made from nankeen that contained rice. They recognized, and suspected, that some of that evidence also suggested close cultural association with the Indians of North America. Great care was taken to restore the grave, retaining only samples of the objects found.[83]

After a visit of only several days, La Pérouse sailed on northward, 27 June 1787, his objective being Sakhalin. Despite published assertions that Sakhalin was a peninsula, La Pérouse believed it to be an island; and, after a visit to Sakhalin, he anticipated proceeding northward through the Tatar Strait en route to Kamchatka. On 12 July 1787, a cove sufficiently sheltered to attempt a landing on Sakhalin was found. The site was named the Baie de Langle as Langle was the first to go ashore.

They were met by a few very friendly natives, about 20 in number, who seemed to be poor, commercial fishermen. They indicated that the cloth for their clothing had come from Manchu country, and they seemed to know about writing from meetings with the Chinese. The French were astonished that the natives pronounced the names *Manchu* and *Sakhalin* exactly as the French did. La Pérouse got one of the natives to trace the outline of Sakhalin in the sand, which confirmed his belief that it was an island; and which portrayed the straits through which they would pass as well as the little stream where they were. No group of people they had met since leaving Brest had so much excited their interest and admiration. The description of the natives' long beards, plus the fact that the landing had occurred along the southern third of the island's west coast, allows the supposition that they had encountered Ainus.

"To find a hunting and fishing people," he wrote, "who do not cultivate the ground in any way, and who do not have any flocks, but who have in general such gentle and serious manners, is very contrary to our accepted ideas. No doubt, the knowledges of the educated class of Europeans far surpasses those of the 21 natives with whom we communicated on all matters at the Baie de Langle. But, the knowledges these people do have seem to be more generally shared within their group than is the case within the common run of the European peoples. All individuals seem to have received the same education. And, in contrast to the dumb incomprehension of the Alaskan Indians at the Baie des Français, our arts and materials attracted the attention of the natives at the Baie de Langle. They

83. Ibid. 3: 13-22.

examined our cloth on both sides, talked about it amongst themselves, and endeavored to discern by what method it had been fabricated. The shuttle is known to them, and I am bringing back a loom on which they make a canvas cloth similar to ours. But, the warp is made from the bark of a common local willow."

If the Ainus lacked agriculture, they did take intelligent advantage of native plants. They had many roots of a lily in their huts that the botanists identified as the yellow lily of Kamchatka, *Lilium camschatcense* L. [=*Fritillaria camschatcensis* (L.) Ker.-Gawl.]. The roots were dried for winter food. They also collected a wild garlic, probably *Allium schoenoprasum* L. var. *sibiricum* (L.) Hartm.; and an angelica along the coast at the edge of the forest, probably *Angelica lucida* L. called wild celery.[84]

Dr. Rollin, who examined the Ainus to the extent they would tolerate, shared La Pérouse's admiration for their intelligence, friendliness, and sense of propriety. While they had no objection to be sketched, they quite refused his attempts to take bodily measurements as if fearful of some magical procedure. He estimated the most common height of the males to be about 5', their build stocky, muscular, with little fat. Their heads were abnormally large, the face larger and rounder than that of most Europeans, animated and pleasant, despite the fact that the component parts lacked the regularity and grace that Europeans take for granted. They had fine teeth in the usual number, but small ears, pierced for wearing beaded ornaments or silver rings. Not only were these islanders heavily-bearded, but their bodies were extraordinarily hairy. The long, full beards gave the older men in particular a serious and venerable air, and the young seemed to have great respect for them.

Their domestic manners and arrangements exposed both refinement and insensibility. Their meals were fairly regular, at noon and evening. Cooking was done in iron cauldrons; eating vessels were mainly wooden; and they used chopsticks in the Chinese fashion. Dr. Rollin would later perceive that Ainu dwellings farther south on the island were somewhat more refined and decorated, more likely to be floored with boards. He would see occasional Japanese porcelain vases, evidently acquired at some expense through trade and highly prized by the Ainu. In general, their wooden huts were well-made and insulated with birchbark. The roofs were a thatch of dry grasses. But the interiors had a stench from fish and fish-oil that the French found intolerable, but which the natives did not seem to notice. In fact, they seemed to flourish in that ambiance. They maintained a good complexion into old age, seemingly healthy, with no sign of contagious or eruptive diseases.

84. Ibid. 3: 26–40.

Each family had its own canoe and equipment for hunting and fishing, using the bow, javelin, and a pike meant for bear. Adjacent to each hut they built storage facilities for winter provisions: dried fish, quantities of wild onions, celery, angelica, and yellow lilies; and the fish-oil they preserved in the stomachs of larger animals they had killed. Whale-hunting seemed to be limited to the extreme south of Sakhalin, and the method of extracting whale oil was hardly efficient. Once the whale had been beached on the talus, it was left to putrify. The oil that naturally seeped from the whale was collected in a tub sunk into the lowest part of the surrounding terrain, where it could receive the oil through small prepared furrows.[85]

The expedition, leaving the Baie de Langle on 14 July 1787, made a second landing to the north on the west coast of Sakhalin on the 19th, a place they named the Baie d'Estaing. There were nearly a dozen native huts nearby. The natives were gathered around four small boats that had been loaded with smoked fish. Their crews proved to be Manchus who had come to trade for fish. During the stopover in Macao, La Pérouse had hired some Chinese sailors as a partial substitute for those crewmen lost in Alaska. These Chinese were now used by Langle as intermediaries with the visiting Manchus, enabling him to have limited conversation with them. They confirmed the fact that Sakhalin was an island, not a peninsula. Previously, these Chinese sailors had not understood a word of the Ainu language.

After distributing presents to both the Ainu and the Manchus, the French returned to their ships, sailing northward (20 July 1787) by following the shoreline. Approaching a remarkable peak on the 22nd, its base down to the water's edge, the ships dropped anchor. The mountain was covered with trees and plants, suggesting a great opportunity for botanical research, and La Pérouse named it Pic Lamartinière. An exploring party was sent ashore, landing in a cove into which a small stream flowed.

The vegetation seemed more vigorous than it had been farther south. Willows and firs were predominant, presumably *Abies sachalinensis* Fr. Schmidt and *Picea ajanensis* (Lindl. & Gord.) Fisch. ex Carr. Probable willows could have been *Salix glauca* L. and *Salix hastata* L. But they also found birch, probably *Betula ermanii* Cham. or *Betula kamtschatica* (Regel) Jansson ex V. Vassil; and a maple, possibly *Acer mandchuricum* Maxim. As for the gooseberries, strawberries, and raspberries they found still in bloom, too many species would have been present on Sakhalin

85 Rollin. 1797. "Dissertation sur les habitans de l'île de Tchoka, et sur les Tartares orientaux." Milet-Murat. *Voyage de La Pérouse*. 4: 73-79.

to justify even a probable identification. In any case, the collections made on 22 July 1787 were highly gratifying.

The following day they proceeded northward along the coast, seeing occasional habitations; but the water was becoming shallower as they moved north, a cause for worry as the sea was increasingly rough and the channel increasingly narrow. On 26 July, La Pérouse ordered a small boat lowered from each frigate to take soundings, each one under the direction of a prudent officer, Lt. Boutin and Ensign Vaujuas. La Pérouse gave their names to the two points that projected into the sea ahead at what appeared to be the narrowest stretch of water between Sakhalin and the mainland (now Lazarev and Pogibi).

After the soundings, and considering the treacherous weather and the frequent heavy fogs, La Pérouse deemed it wise to seek an anchorage on the mainland. A cove was found on 28 July 1787 that he named Baie de Castries, in the proximity of a Tartar village. The prudent pause gave opportunity to ponder the difficult situation. He recognized that, if he dared not go farther to pass Sakhalin, he would have to backtrack against winds that were still from the south and might not reverse until November. That implied that he might not be able to reach Kamchatka in the good season.[86]

Consequently, La Pérouse resolved to remain no longer than 5 days, time to take on wood and water and to give the naturalists opportunity to make collections and to observe the inhabitants, who called themselves Orochis. Like the Ainu of Sakhalin, these Tartars were primarily dependent upon the abundant salmon, although they also had some grain obtained through trade with Manchus to the south. And, like the Ainu, they showed great respect for individual property. Nothing was stolen from the French. The roots of the yellow lily were collected and dried for winter use; but, as they had no agriculture or flocks, it was presumed they could not advance. They were seen to eat raw the nose, gills, and sometimes the entire skin of a salmon, a disgusting primitivity from the French viewpoint.

Yet, they appeared to be a very gentle people, respecting the elderly and devoted to their children. They seemed to recognize no chief or have any visible government, and no priests or temples were evident: an apparent anarchy without any quarreling. Those glimpses of an Edenic people were again spoiled by a pervasive smell of fish. Indeed, the Tartars were very unclean, emitting a revolting odor, so very far removed from a European ideal of beauty. The women appeared to be hardly distinguishable from the beardless men, identifiable primarily by dress. It was noticed that in trade, no transaction took place without the consent of the

86. La Pérouse. 1797. *Journal.* 3: 40-57.

women. Copper ornaments were reserved for women and girls alone. From the great consideration the women evidently enjoyed, one might deduce that government did exist, if one not recognized by Frenchmen.

The naturalists remained active during that late-July visit. Lamartinière searched in vain for new plants but found nothing he had not already seen at Ternej or on Sakhalin. Strawberries and raspberries were still in flower, but the gooseberries were beginning to redden. An edible celery or cress was quite rare. Collignon did plant European seeds in the neighboring woods (we are not told what they were), but probably without much confidence of a happy outcome given the cultural level of the natives.[87]

Despite the little variation in their diet, Dr. Rollin found the younger Tartars to be in very good health. But, as they advanced in age, they seemed to be increasingly subject to conjunctival inflammation and to blindness. The frequency of such ailments suggested general causes: perhaps the glare from the snow that covered the ground for more than half of the year; perhaps the exposure to smoke in their huts during cold weather or to the smoke used to dispel the swarms of mosquitoes during the summer months. Skin diseases seemed to be rare even though the Tartars lived in extreme filth. He saw no trace of smallpox or syphilis.[88]

La Pérouse did depart as planned on 2 August 1787, turning southward along the mainland coast, constantly tacking to cope with the prevailing south wind. When he reached the southern tip of Sakhalin, on the strait separating Sakhalin from Hokkaido, he named it Cap Crillon, today La Pérouse Strait. An Ainu vocabulary that had been prepared by Dr. Lavaux, chief-surgeon on the *Astrolabe*, was pressed into service at Crillon and helped to allay initial suspicions. For it was apparent to the French that those Ainu in the south, closer to Hakkaido, had had considerably more transactions with foreigners than the more northerly Ainu, and probably had good reason to regard them with suspicion. Once the French succeeded in gaining their confidence, they came on board the two frigates as if visiting old friends, sitting in circles and smoking their pipes. Since leaving Manila, this was the first time any natives had shown any interest in visiting the ships.

They were unquestionably Ainu: well-formed, vigorous men, beards extending on to their chests, their arms, necks, and backs also covered with hair. They were simply much more advanced in their industries; their habitations had a considerable degree of cleanliness and refinement; and they had furnishings of artistic Japanese manufacture. Fishing and hunting still provided the basis for subsistence, but their production of whale oil had given them commercial advantages.

87. Ibid. 3: 57-78.
88. Milet-Mureau. 1797. *Voyage de La Pérouse*. 4: 80-85.

Despite their advancement, available fertile land was uncultivated, and they had no flocks.

The French made many gifts to them for which they were expressively grateful. La Pérouse soon perceived that what the natives most wanted was eau-de-vie and tobacco, commodities he allowed to be distributed very sparingly. He needed the tobacco for his crewmen, and he greatly feared the consequences of the eau-de-vie, and was reluctant to provide it at all. The Ainu, though they had salmon in their boats, offered nothing in return, preferring to sell it at prices so excessive that the French refused to buy. In any case, La Pérouse had already recognized the moral implications of buying either dried or smoked salmon from either the Ainu or the Tartars. They should not be tempted to sell what very well could be their supplies for the winter, opening them to starvation in that season.[89]

After a short anchorage, the expedition sailed eastward, 15 August 1787, through La Pérouse Strait toward the Kurile Islands, a chain originally explored by the Dutch in the 17th century. Once they reached the chain and turned northeastward, the lack of anchorages, the dense fogs, and the violent currents soon deflated all enthusiasm for exploration and led La Pérouse to aim directly for Kamchatka. Its coast came into view on 5 September 1787, a forbidding sight. Two days later, they entered Avacha Bay to anchor at Petropavlovsk. The Russians had been informed of the probable visit, and La Pérouse had arrived equipped with a Russian interpreter.

Jean-Baptiste-Barthélemy, baron de Lesseps, had been born to a family traditionally in the diplomatic service. His father, Martin, had been the French consul-general in St. Petersburg where the young son obtained a thorough knowledge of the Russian language. That accounted for Louis XVI's personal decision to add him to La Pérouse's staff with the rank of ensign. Finally, at Petropavlovsk, he could fulfill his assignment.

From Petropavlovsk, the naturalists could see a volcano at no great distance, still snow-covered and obviously still active: part of a chain of volcanoe that run from Cape Lopatka to Cape Kronoki, and presumably Mt. Korjakskaja. The commander of the local garrison, Captain Schmalov, assigned eight Cossacks to accompany scientists meaning to explore the mountain. They recognized the probable difficulty of the ascent and planned a four-day trip. It began easily enough through a gentle terrain covered with plants, many of them berry-covered. They noted stunted, nearly dwarf firs; some birch; and a pine with small nuts good to eat, probably *Pinus pumila* (Pall.) Regel. Only on the afternoon of the third day did they reach the top of the crater, hampered by deep snow and blinding fogs.

89. La Pérouse. 1797. *Journal.* 3: 79-92, 110, 116.

Meanwhile, the two frigates were taking on wood and water, but no cattle could be purchased. Despite having fine meadows, the native population, the Kamchadales, had long since given up with cattle, defeated by winters seven to eight months long, preferring to rely on hunting and salmon-fishing for subsistence. Most of the Russians and Cossacks present, soldiers rather than cultivators, had simply adapted to the native regime. Only Captain Schmalov, his Lieutenant Kaborov, and their sergeant maintained gardens, rather pathetic outposts of European life, full of potatoes and turnips. Their exhortations for emulation had been quite fruitless. The locals preferred what they could pull up in the wild: mostly the yellow lily, *Fritillaria camchatcensis* (L.) Ker-Gawl.; and wild garlic, *Allium schoenprasum* L.; plus berries they made into beverages or preserves. Consequently, Collignon gave a substantial quantity of seeds to the three gardening officers, and to them alone, with the understanding that they would send him reports on the results of the next spring plantings.

As the governor of the Maritime Province, Colonel Kaslov-Ougrenin, was known to be on a tour of the province from his seat in Okhotsk, his arrival in Petropavlovsk was immensely anticipated by the French on the assumption that he would bring letters from France. The letters, in fact, arrived only several days later by courier from Okhotsk, and the governor had managed to assemble seven beef cattle that he insisted be accepted as a gift from the Russian government.

La Pérouse obtained the governor's permission to send a second portion of his journal across Rusian territory to France. The explanation La Pérouse gave for his decision to have Lesseps carry the letters, maps, and journal to France (that, as a future French diplomat in Russia, it would be a service to Lesseps to acquaint him with the Russian provinces) is not quite convincing. La Pérouse was aware that he enjoyed the special confidence of Louis XVI; he knew that Lesseps had been personally selected by the king for the expedition; so that Lesseps remained the ideal conveyor of either delicate or confidential information La Pérouse might want to transmit directly to His Majesty. Otherwise, the packages could have been sent through the Russian post as La Pérouse knew.[90]

Once these arrangements were completed, and Lesseps had become the charge of Governor Kaslov, the French expedition, aware of the oncoming winter, sailed southward on 29 September 1787. The separation was a very difficult and emotional moment for all concerned. No explanation has been offered for why, in view

90. Ibid. 3: 94, 99-101, 126-155.

of impending winter, the governor sent Lesseps to Okhotsk by land rather than by the much shorter route by sea. In any case, a caravan of 35 dogsleds, later drawn by reindeer, left Petropavlovsk on 7 October 1787. As described by Lesseps, the Siberian cold, snow, and wind made the trip almost unendurable and seemingly endless.

The party did not reach Okhotsk until 5 May 1788 after nearly seven months of what was clearly a struggle to survive. From Okhotsk, the trip across Asia would be accomplished in better weather and by utilizing the great rivers of Siberia. He entered St. Petersburg during the night of 22 September 1788. Following his instructions from La Pérouse, Lesseps delivered his packages to Louis-Philippe, comte de Ségur, minister plenipotentiary to the empress, for transshipment to the ministry of the marine. After only four days of rest, he proceeded to Versailles, which he reached on 17 October 1788 and was received that very day by the king.[91] A year had gone by since he had separated from the expedition.

Meanwhile, that expedition had sailed southward along the east side of the Kuriles chain, the weather too bad and the swells too heavy to risk any exploration. On 5 November 1787, they crossed the line that they had earlier followed from Monterey to Macao, no land having been sighted for weeks. Nor were they catching fish as anticipated, the resultant hunger contributing to the evident fatigue of the crews. Their luck changed the following day when 8 tuna were caught, providing a nourishing meal for officers and crew, and the weather began to improve. The capture of several large sharks a few days later also augmented the supply of fresh food. But La Pérouse was becoming alarmed by the failure to encounter islands reported, or mischarted, by earlier voyagers, and concluded that he must set a course for the island group Bougainville had named the Iles des Navigateurs, which we know as the Samoa Islands.

They first approached the eastern end of the Samoan chain on 6 December 1787. It was difficult to find a suitable anchorage as the islands were bristling with outer reefs, and the swell was high. Islanders came out in some numbers in pirogues and were unarmed, suggesting they were peaceful. Some trading took place, the natives preferring glass beads to iron objects. A wooden pot, filled with coconut oil they offered for sale, testified to considerable industrial accomplishment, as did their artistically constructed pirogues.

Sailing on westward toward a larger island indicated by Bougainville, they approached Tutuila Island on 9 December 1787. Reefs were again a threat, but the pirogues came out easily through passages between reefs, indicating that French

91. Jean-Baptiste-Barthélemy, baron de Lesseps. 1880. *Voyage de M. de Lesseps du Kamtchatka en France*. Paris: M. Dreyfous, pp. v-vii, 7-8, 74, 186, 233-235.

longboats could pass through similarly to shore. Even though trading for pigs, coconuts, and other fruits began offshore, Captain de Langle and several officers took three armed boats ashore to pursue trading for food, remaining for about one hour. They met a friendly reception.

As the following morning was a beautiful day, La Pérouse thought they should take advantage of the fine weather to go ashore, both to take on fresh water and to study the inhabitants. But it was understood that they would sail before dark to avoid spending a second night at their risky anchorage. Some crewmen were sent ashore to obtain water. La Pérouse and Langle followed in separate boats to reconnoiter. Once on shore, a detachment of soldiers formed a protective barrier between the boats and the natives.

From the very beginning of the expedition, La Pérouse had always taken military precautions when in the presence of aboriginal populations. While he was sympathetic to them, he was also quite unhampered by any fashionable theories in their regard, quite convinced on the contrary of the blessings that come from living in civilization.[92]

As the water casks were being filled without any trouble, La Pérouse felt free to go several hundred steps to visit a charming village set in a grove of fruit trees. Invited to enter a chief's dwelling, he found it surprisingly well furnished and decorated. "This delightful country," he recorded, "still combined the double advantage of a fertile land requiring no cultivation with a climate that required no clothes. Breadfruit trees, coconut palms, banana trees, guavas, and oranges offered these fortunate people healthful and abundant nourishment. Chickens, pigs, and dogs, living off the surplus of these fruits, offered them a pleasant variety of dishes. They were so wealthy, they had so few needs, that they distained our iron instruments and our cloth, valuing only our colored-glass trinkets and beads. Quite fulfilled by real blessings, they desire only the useless."

From which, he recognized, it might to assumed that the Samoans must be the happiest people on earth, their days pure and tranquil; and that would be deceptive. "Although we saw no armament, their bodies bore the scars of frequent fighting, whether against others or amongst themselves. The men's features displayed real ferocity. Without doubt, Nature has left this imprint on the faces of these Indians as a warning that the nearly savage man, living in anarchy, is a being more wicked than the most ferocious animals."[93]

92. Gaziello. 1984. *L'Expédition de La Pérouse*, p. 197.
93. La Pérouse. 1797. *Journal*. 3: 165-192.

La Pérouse, himself, had experienced no impropriety during his inspection of the village, but thereafter he began to hear of disturbing incidents. A native had thrown stones at Dr. Rollin, the chief surgeon. Another had attempted to snatch Captain Manneron's sword, revealing that the men were unruly and quite undisciplined by their chiefs. What is more, the men, in physique, were more impressive than any native group the French had seen during the voyage: taller, 5' 9 to ll", and powerfully built, equal or superior in strength to Europeans. Any dispute amongst them quickly led to blows with sticks, clubs, or paddles.

As for the women, he found them also tall and with very good figures, at least briefly, but lost their elegant form while still quite young. The suggestive movements by a majority of the women, which La Pérouse found indecent, the "repugnant offers" they made for their favors, fixed the attention of some of the crewmen. He felt obliged to report that, despite his prohibition, liaisons were arranged, primarily negotiated by older women. It appeared to La Pérouse that the young women who submitted were all unmarried and that such conduct was forbidden to married women. Copulation was carried on as an organized ceremony, an elderly man presiding in a hut with elderly women singing and howling as the background. The entire performance, in fact, suggested a commercial prostitution available to outsiders, not practices condoned within the tribe.[94]

Everything La Pérouse had seen on shore, 10 December 1787, strengthened his earlier resolve to leave that anchorage offTutuila before nightfall. He returned to his ship shortly after noon with the other boats close behind. They found if difficult to get back on board because of the throng of native pirogues. Once on board, La Pérouse found that Lt. Boutin, who had been left in command, had been unable to prevent some Samoans from climbing aboard; and he had not fired upon them as he was aware of La Pérouse's principles. It had become evident that the natives believed they had nothing to fear from French weapons. La Pérouse at once ordered the sails to be raised. The natives plunged overboard as soon as they recognized they were being carried away.

Hardly were the frigates underway when Captain de Langle put in an appearance to describe a superb bay westward down the beach he had reconnoitered while on shore. He had found a cascade of exceptionally pure water flowing into the sea at that point. As Langle subscribed to Captain Cook's conviction that scurvy was the product of a deteriorating water supply, and as he had several crewmen on his ship who were threatened by scurvy, he wanted to put several boats on shore in

94. Ibid. 3: 225-228.

that bay to secure additional fresh water, promising to return to the frigates within three hours.

As it was getting late in the day, the frigates stood out to sea for the night. The plan was for Langle to go ashore next morning, 11 December 1787, with two longboats, one from each frigate. The two parties came to a total of 61, including the naturalists Lamanon, Lamartinière, Father Receveur, and Dr. Lavaux, who took advantage of even a brief opportunity to collect. Everyone, all 61, was armed with guns and sabres, no matter that the inevitable natives in pirogues seemed entirely peaceful.

As the two parties approached Langle's fine bay, they were dismayed to find it full of coral reefs. He had obviously seen the bay at high tide the previous day, and the tide had now ebbed. As they struggled to get to shore through water that was too shallow, the Samoans soon perceived that the French were in difficult circumstances, and the stones began to fly. Langle was apparently reluctant to give the order to fire, perhaps thinking he could extricate the parties without bloodshed. He paid with his life for his humanity. Knocked overboard by flying stones, he was immediately murdered by the natives.

In the subsequent melee, 49 members of the expedition managed to escape, many of them badly injured. Both Langle and Lamanon were among the 12 killed outright, and Father Receveur would succumb from his wounds a few weeks later at Botany Bay. One of the longboats had also been lost. Nearly 100 native pirogues were still milling in the vicinity of the frigates when the survivors arrived, and it was all the angry La Pérouse could do to keep his crews from firing on them. One cannon was fired to warn them to leave.

La Pérouse ordered Ensign Vaujuas, who had taken command of the retreat from the bay, to write down his version of the events of 11 December 1787 to clarify what had been observed from shipboard. La Pérouse pondered a considerable time about whether he should seek revenge for the unprovoked attack, concluding that it would be better to proceed to Botany Bay where a new longboat could be built. A commander less conscientious than La Pérouse could have found others to blame for the disaster. Instead, he confessed that, though he had had misgivings about Langle's proposal to go ashore for the pure water, he had nevertheless given his consent. It was an admission of bad judgment and an assumption of responsibility. He portrayed Captain de Langle as an officer concerned for the health of his crew, a man who had lost his life in what turned out to be a vain attempt to avoid bloodshed.

As they sailed westward through other Samoan islands, these events weighed heavily on his spirit: "As for the morality of these people, although we saw them only briefly, we learned only too well about their character through our misfortunes; and we do not fear to say definitively that you will seek in vain to arouse any gratitude in these ferocious souls through good deeds, as they are constrained by fear alone."[95] He would still be brooding about the incident when they reached Botany Bay.

Despite what had been learned about the danger of going ashore on a Samoan island, Lamartinière and Collignon did subsequently land on several islands to botanize. The natives demanded a glass bead for every plant Lamartinière collected. On one occasion they threatened to club him when he declined to pay. They went alone. La Pérouse refused to send any crewmen ashore unarmed; and he feared that armed sailors would seek revenge against the natives, so great was their indignation about the massacre. If botanomania had to be served, the Navy would only watch and wait.

La Pérouse would not even anchor close to shore for the remainder of the trip to Botany Bay. He had the mapping limited to charting routes around the islands. Natives did come out frequently to exchange goods, but only at some distance from land. Earlier benevolence now gave way to rigorous discipline: no native was welcome to come on board, and any native who attempted to board in defiance of the prohibition was threatened with death. Both reason and common sense had driven La Pérouse to regard every native as a potential murderer. Without regret, the expedition left Samoan waters on 23 December 1787.[96]

A southwestern course took them toward the Tonga Islands, then known as the Friendly Islands. They sighted Vavao Island (Vava'u) on the 27th, but the sea was extremely rough and no pirogues came out from shore. They saw no natives until they reached Tongatapu Island on 31 December 1787. The contrast with Samoa was immense. The entire island seemed to be cultivated; and, in that rainy season, it was very green. The natives' huts were not grouped in villages but were scattered among the fields. The people were farmers, and better farmers than sailors in view of their noticeably fragile pirogues. They brought out coconuts and bananas to trade, but no chickens or pigs as their pirogues could not support such cargo in the heavy groundswells. They traded in very good faith for iron, nails, and small axes. A considerable mutual confidence developed, and La Pérouse had no objection to letting them come on board.

If obviously akin to the Samoans in ancestry, the Tongans were physically inferior and lacked any ferocity in their features. La Pérouse presumed the physical inferiority reflected the arid, less fertile environment of the Tonga archipelago, but also suspected they owed their greater progress in civilization to their need

95. Ibid. 3: 192-225. Also see Dunmore. 1965. *French Explorers in the Pacific.* 1: 281.
96. Ibid. 3: 235-241.

to develop agriculture for survival. He saw their development in several arts as compensation for want of brute strength. Indeed, the only weapon he saw them carry was a patu, a short club with sharpened edges. It was found to weigh less than a third of the clubs they had acquired on Tutuila.

It became quickly apparent that the expedition could not obtain sufficient provisions on Tongatapu to justify an extended visit. They set out again for Botany Bay the next day, 1 January 1788. On the 13th, they sighted the cliff-bound coast of Norfolk Island. As the botanists had had so little opportunity for extended research since Kamchatka, La Pérouse meant to send them ashore. But the breakers proved to be too rough for a safe landing.[97]

The expedition reached Botany Bay on 26 January 1788. They found a British fleet already at anchor, having arrived on the 18th with Governor Arthur Phillip on board. It was a welcome sight for, as La Pérouse put it, "all Europeans are compatriots at this distance from home." While the British offered what services they could, the fleet was on the verge of sailing north to raise the flag in Sydney Cove.[98] In fact, the primary need at that moment was wood and water, which they knew would be readily available, plus the time to build a longboat to replace the one lost on Tutuila.

So ended the personal narrative of La Pérouse.[99] While the reader knows that letters and other materials had been sent to France from Macao in January of 1787, the first segment of the journal was actually mailed from Manila; Lesseps had carried the second section from Kamchatka; and La Pérouse would now send the third part from Botany Bay.[100] The few illustrations of native peoples that accompanied these shipments, and would be published by Milet-Mureau in his atlas, cannot be regarded as reliable records. They were highly idealized and not scientific drawings, reflecting contemporary neoclassicism. It has been proposed, even so, that the drawings did reveal a change of attitude in the French: from respect to distrust of native character.[101]

La Pérouse quite understood that his journal would be treated as an official report to the Navy and might well be published. Consequently, his description of the massacre of 11 December 1787 had been truthful, but not the whole truth, omitting critical details potentially embarrassing to the regime should they become

97. Ibid. 3: 242-262.

98. J. H. Maiden. 1910. "Records of the Earlier French Botanists as Regards Australian Plants." *Journal of the Proceedings of the Royal Society of New South Wales*. 44: 125.

99. La Pérouse. 1797. *Journal*. 3: 264-266.

100. La Pérouse to the Minister of the Marine, 5 February 1788. Milet-Mureau. 1797. *Voyage de La Pérouse*. 4: 200.

101. Smith. 1985. *European Vision and the South Pacific*, 1768-1850, pp. 139-141.

known to opponents of such expeditions or to the subscribers to noble savagery. He entrusted such details to an *unofficial* letter to the minister: After the murder of Captain de Langle, an order *had* been given to fire in self-defense. The crewmen in the longboats had then killed at least 30 Samoans in the skirmish, allowing a majority of the French to escape. What is more, at the moment of that unanticipated attack, which had been quite visible from the frigates, hundreds of Samoans were around the ships engaged in trading. The infuriated French crewmen, had they not been restrained by the ships' officers at La Pérouse's order, could have killed another 500 natives.

At Botany Bay, as the British ships had by then moved on, he assured the minister that he was constructing a palisade on shore to provide security while new longboats could be built, a task that would probably occupy the rest of February. He would leave nothing to chance. His view of uncivilized peoples had been formed some time ago, probably in Canada, and the voyage of the moment had only served to confirm it. He was aware that the local Australian aborigines, if weak in number and unequally armed, could be very malicious: "After accepting our presents and our expressions of friendship, they threw spears at us. Even so, I am a thousand times angrier at the philosophes, who exalt the savages so highly, than I am at the natives themselves.

"The unfortunate Lamanon, whom they murdered, said to me on the very eve of his death, *those people are said to be worth more than we are.* As a strict observer of the orders written down in my instructions, I have always treated the natives with the greatest moderation; but I swear to you that, if I were ever again to undertake another campaign of this sort, I would ask for different orders. A navigator leaving Europe must regard aborigines as enemies: in truth very weak, whom it would be ignoble to attack without motive, and whom it would be barbarous to destroy. But who you have the right to forestall when legitimate suspicions authorize you to act."[102] He confessed that the loss of Langle and the events on Samoa still weighed heavily upon him despite efforts to put that day behind him.

Finally, La Pérouse thought it well to describe the route he proposed to take after leaving Botany Bay: He would backtrack to the Tonga Islands, going by way of the southern part of New Caledonia. Then he would sail northward to Santa Cruz Island (Nendo Island), the Solomon Islands, and the Louisiade Archipelago. He would attempt to determine whether the latter was a part of (or separate from) New Guinea. By the end of July 1788, he expected to be passing between New

102 La Pérouse to the Minister of the Marine, 7 February 1788. Milet-Mureau. 1797. *Voyage de La Pérouse.* 4: 233-241.

Guinea and New Holland (Australia) in order to visit the Gulf of Carpentaria in September; then down the west coast of Australia and along the south coast as far as Tasmania. He would cut short that portion of the trip, however, in order to reach Mauritius by the beginning of December 1788.[103]

As the itinerary did not specify any European outpost as a port of call, it implied no further communication for roughly ten months. In fact, after La Pérouse sailed from Botany Bay on 10 March 1788, he was never heard from again; and it would necessarily be many months before Paris could even begin to suspect a disaster at sea.

103 Ibid. 4: 202-203.

VI
Adventures and Misadventures: The Bizarre Story of the Dupetit-Thouars Brothers

The year 1789 passed without a word from La Pérouse and without any report of wreckage sighted. As the months went by, the probability of shipwreck had to be recognized within the ministry of the marine; but one could hope for the possibility of survivors ashore on some remote island living in expectation of rescue by compassionate compatriots. The mobilization of any such effort by official Paris during the confused opening months of the Revolution was obviously out of the question, leaving the way open, at least for the moment, for private individuals to take an initiative that was a mixture of audacity and madness. Indeed, their plan, hatched in 1790, proved to be so costly to organize in both money and time that the belated official rescue expedition had finally left port some months before the private venture put out to sea.

The enterprise had originally been the dream of Aristide Aubert Dupetit-Thouars. He was the younger brother, born to a family traditionally in military service whose seat was the château de Boumois in the commune of St Martin-de-la-Place, just north of Saumur across the Loire. He had entered naval service in 1778 and saw naval action during the American Revolution.

His older brother, Louis-Marie-Aubert Aubert Dupetit-Thouars, also in military service, but an amateur botanist, described his younger brother as having a lively mind and a fervent imagination. Naval action had fed his zest for activity, but the subsequent years of peace had deprived him of that nourishment. The image of La Pérouse, waiting in vain on some rock to be rescued by compatriots, a man whose love for glory and humanity he shared intensely, inflamed his imagination and his passion for a voyage of discovery.

Various projects formed in his mind successively until he decided that the most appropriate way to promote his goal would be to open a public subscription to provide for the outfitting of a ship that could undertake a search for La Pérouse. Although he hoped that others would share his humanitarian enthusiasm, he had come to think it prudent to add the enticement of financial reward. During the voyage, the ship would engage in a trade for furs on the northwest coast of North

America, and subscribers could realize considerable gain from their investment. An irresistible package: profits with honor. His ideas were set forth with great enthusiasm in several prospectuses.[104]

Aubert, the older brother, had entered the military service in 1775 at the age of 17. He tells us that his study of botany began in 1780, and that it had been necessary to botanize wherever his regiment was garrisoned in France. But when he was on leave, his favored region was his home country in the Loire valley, around Saumur, le Petit-Thouars, Chinon, and Thouars; and he is known to have been an affiliate of the Société des botanistes d'Angers, an active group of botanophiles.[105]

As he himself recalled, as soon as a young man has tasted the charms of botanical science, he experiences a kind of eccentricity that draws him to more distant paths. He requires new nourishment as he exhausts neighboring gardens and enclosures. Since nearby regions had all been scoured by many predecessors, he can only hope for gratifying discoveries at a great distance. After casting his eyes over the globe with the aid of maps, his taste, or rather his passion, for lengthy voyages is awakened.

Once Aubert knew of his brother's project, he saw it as the most favorable opportunity he would likely meet. "I at once tied my destiny to his, assuring him that, if he were to be the Cook or the Bougainville of the venture, I would be its Banks or Commerson." Surely the fraternal bond would unite the two aspects of the voyage, producing the harmony necessary for the progress of science that had been lacking on prior great expeditions.[106]

Consequently, he resigned from the service to devote himself completely to preparation for the voyage to seek La Pérouse. Aristide wrote to the minister of the marine on 11 January 1791 to announce the intentions of their expedition in the expectation of governmental financial support. He meant to acquire a brig, the smallest feasible ship: light, maneuverable, and requiring only a small crew. The furs that were to be obtained in Alaska would be taken to China for profitable sale.

By then, the ministry was planning a major expedition of its own to search for La Pérouse and could not entertain any diversion of funds. Because Aristide was a lieutenant in the royal navy, the National Assembly finally voted him a subsidy

104. L.-M.-Aubert Aubert du Petit-Thouars. 1804. *Histoire des végétaux recueillis sure les îles de France, La Réunion (Bourbon) et Madagascar*. Paris: Mme Huzard, pp. vi-vii.

105. Alexandre Boreau. 1845. "Notes sur la vie et les travaux du botaniste Aubert du Petit-Thouars." *Bulletin de la Société industrielle d'Angers*. 16: 55-57.

106. Aubert du Petit-Thouars. 1804. *Histoire des végétaux*, pp. v-vii.

of 10,000 livres on 22 December 1791 (by which time the official expedition was already at sea) and guaranteed his naval pay for two years while at sea.[107]

The failure to obtain more than token official assistance was the more serious in that the support of a few early subscribers was not sustained. Even worse, Aristide's lack of experience made him a dupe in those markets where he was required to sign obligations, so that outfitting expenses considerably surpassed his anticipations. The brothers were soon forced to sell their small inheritances, which went entirely into the expedition; and ultimately their brothers and sisters came to their aid with sums that were considerable in view of the modesty of their means. Such combinations enabled Aristide to purchase a small ship, *Le Diligent*, in Rouen, to complete its outfitting in Le Havre, and to proceed to Brest where he was to meet Aubert.

The latter, taking advantage of the opportunity to botanize at leisure as he crossed Brittany toward Brest, soon ran afoul of local revolutionary committees (it was 1792), which were on the alert for émigrés and other possible counter-revolutionaries. It becomes suddenly apparent that these two brothers, so entirely consumed by their passions to find La Pérouse and to botanize in faraway places, had been untouched by the political turmoil of their times. Indeed, Aubert was so innocent of any political insight that he could not have imagined that a wandering botanist, putting plants in a box, was bound to arouse the suspicions of village patriots. He was placed under arrest, and the magistrate who interrogated him no doubt thought he was saving the nation when he placed seals on Aubert's collection box. In the end, he was acquitted of the charges in the indictment, but only after six weeks of detention.

That delay profoundly altered the course of his life. When he reached Brest, he found only a message and directions from Aristide. From it, he learned that Aristide, in Brest, had soon found himself the object of suspicion. As he became aware of absurd rumors about him being spread by local revolutionary groups, he began to fear that his life could be in danger if he did not put out to sea. As he could not account for Aubert's failure to appear, he put all his equipment on board; and, after leaving word that he would await Aubert in Mauritius, Aristide had departed on 22 August 1792.[108]

Details of his fate are few. It would appear that some of his crew of 32 were infected unknowingly during a call in the Cape Verde Islands. A violent fever broke out after they were again at sea. Two-thirds of his crew was ill, and his

107. Richard. 1986. *Le Voyage d'Entrecasteaux*, pp. 25-27.

108. Aubert du Petit-Thouars. 1804. *Histoire des végétaux*, pp. vii-viii; Boreau. 1845. "Notes sur la vie and les travaux de botaniste Aubert du Petit-Thouars", pp. 58-59.

surgeon died. He sought help at the Portuguese island of Fernando de Noronha off Brazil, only to be greeted as an enemy and imprisoned. In due course, the survivors and their small ship were escorted to Lisbon, from where Aristide was able to notify the ministry of the marine of his dilemma.

A French naval officer was then sent to Lisbon to negotiate his release in August of 1793, after which he was able to sell his ship and pay off his crewmen. Having no furs, he could not pay off his subscribers. His ill-fortune had not yet come to an end. We last hear of him in command of the battleship *Tonnant* in Abukir harbor during Bonaparte's invasion of Egypt. Nelson surprised the French fleet at anchor and disposed of it rather easily on 1 August 1798. Captain Dupetit-Thouars, with one leg broken and a foot blown off, had continued to shout orders from the deck until his ship sank, carrying him with it.[109]

But what of Aubert Dupetit-Thouars, stranded in Brest in September of 1792? Since all his possessions had been put on board *Le Diligent*, he had little option except to seek passage on the next ship leaving Brest for Mauritius. Accordingly, he embarked on a small ship on 2 October 1792, which only had a crew of nine; and he would later confess that he would never have subjected himself to such a frightening prospect had his circumstances been less desperate.

The principal dilemma they faced during the passage to Cape Town was not a rough sea, but rather six weeks of calm in the doldrums south of the equator. The ocean currents pushed them considerably to the west of their course during that period, and they were unable to make any headway. Faced with a growing shortage of water, the captain knew he must land on the first available island for a fresh supply. The first island sighted proved to be the principal island in a small group, Tristan da Cunha, with its highest peak shrouded in clouds. It was 2 January 1793, exactly four months after the departure from Brest. They maneuvered all night in order to remain close to the islands.

At dawn on the 3rd, they found themselves close to the principal island, but also could see a smaller island to the south. Despite its name, Inaccessible, they coasted it in seach of a landing place. Finding nothing suitable, they returned to Tristan da Cunha, anchoring about noon on the north side of the island. Some of the crew immediately dropped fishing lines into the water and began bringing up a species of wrasse in particular. Dupetit-Thouars went ashore about 2 p.m.,

109. Richard. 1986. *Le Voyage d'Entrecasteaux*, p. 27; Horner. 1995. *Looking for La Pérouse*, pp. 10-11; Gabriel-Alexandre Marcel. 1888. *Une Expédition oubliée à la recherche de La Pérouse*. Paris: L. Baudoin, is a brief sketch of the Dupetit-Thouars expedition.

re-embarking at 7 o'clock: "It is easy to imagine how impatient I was to get ashore and to take possession of that island in the name of Flora."[110]

Penetration of that new land was a step-by-step battle. He met areas densely occupied by a reed-like grass, presumably a *Phalaris*. Along the banks of a stream, which formed a cascade that emptied into the bay, ferns and maidenhair formed a dense carpet. Other areas were covered with a shrub in the genus *Phylica* L. [ex Rhamnaceae, the Buckthorn Family]. Its stalks, crammed together and interlaced, perishing only from old age, provided a nearly impenetrable barrier. Only after extreme exertion did he return to the ship with about 50 plants, most of them appearing to him to be new.

The following day, 4 January 1793, the ship's boat could not go ashore as it required repairs to its hull. He took advantage of the opportunity to examine his collection. Never before had he made a satisfactory effort to draw plants. Necessity now made him more skillful, and thereafter he would always sketch those plants that seemed most important to him. On the 5th, he went ashore at noon, returning at 8 o'clock with about 30 plants he had not previously seen.

Because of the small size of the crew, which was necessarily engaged in refitting the ship in preparation for departure, he could not call upon any crewmen for assistance in his research, and his time on shore was dependent on their availability. They did land him at 9 o'clock on the 6th, when he undertook to climb to the summit of the mountain, a volcanic cone 7,640' in elevation. Although he reached the top, the difficulties of both the ascent and descent were quite beyond his expectations, and the onset of darkness and rain obliged him to take shelter for the night beneath a Phylica. Hunger and cold prevented him from sleeping, and he realized that the sudden onset of bad weather might force the captain to weigh anchor and leave. This was not a voyage, after all, dedicated to the progress of science, and Dupetit-Thouars spent the night with visions of becoming a new Robinson Crusoe.

When he regained the ship at about 10 o'clock the following morning, with only 20 new plants as the reward for his ordeal, he found that his disappearance had produced a gratifying alarm for his safety on board. He also learned the disconcerting truth that the captain, assuming he had gone astray for good, had been inclined to leave his effects on shore for him and to proceed with the voyage. After receiving kindly rebukes for the anxiety he had caused, he joined in preparations for immediate departure, 7 January 1793.

110. Aubert Dupetit-Thouars. 1811. "Description abrégée des isles de Tristan d'Acugna; lue dans la Séance particulière de la première Classe de l'Institut, en janvier 1803." *Mélanges de botanique et voyages*. Paris: Arthus Bertrand, pp. 4-6.

At first, they coasted along the east side of the island. The captain again sent his boat ashore in the hope of finding turtles. None were found, and they had to settle for filling the boat with penguins, which allowed themselves to be taken easily. They would provide an important food supply during the 26-day voyage to the Cape of Good Hope. The birds refused to take any food, but splashed around from time to time in kettles of seawater provided them and ran freely between decks.

Dupetit-Thouars found no new plants on that east side of the island to add to his herbarium, leading him to conclude that he had probably already collected the major part of the island's vegetal wealth. The material he had gathered on the island provided him with a pleasant occupation until they reached the Cape of Good Hope; but, deprived of the library he had assembled for the voyage with his brother, he had only one reference book, Jussieu's *Genera Plantarum*. Although he would be able to borrow additional volumes for consultation once he reached Mauritius, the loss of his own library would hinder him in making plant descriptions with the desired exactitude for the remainder of his years abroad, not infrequently bringing his research to a halt.[111]

He had collected roughly 100 species on the parts of Tristan da Cunha he covered, among which there were only 14 dicotyledons. The most noticeable among them was the tall shrub, growing to 15' in height, that he named *Phylica arborea* Thou. [=*Phylica nitida* Lam.], a buckthorn. It was the only plant on the island that could provide firewood or be suitable for ordinary construction. There was also a low shrub, an *Empetrum*, that he took to be *E. nigrum* L, the crowberry. It had a small reddish fruit with a tart, quite pleasant taste.

He found two aromatic plants: a chenopod with tomentose leaves, different from both *Chenopodium ambrosioides* L. or *Chenopodium botrys* L., which he named *Chenopodium tomentosum* Thou.; and a geranium, or rather *Pelargonium*, which had some affinity with *Pelargonium odoratissimum* (L.) L'Hér. ex Ait., but seemed to be different. He named it *Pelargonium acugnaticum* Thou. [=*Pelargonium australe* Jacq.], claiming it to be the most attractive plant in the flora of Tristan da Cunha.

The only plant he found suitable for cooking was a wild celery he named *Apium australe* Thou. [=*Apium prostratum* Labill.]. (Curiously enough, both authors had been on expeditions to find La Pérouse.) The celery had a strong taste, was somewhat aromatic, but gave no hint of anything poisonous. His example did not convince any of the sailors on board that it was safe to eat, but

111. Ibid. pp. 6-8.

he suspected that it might be the same species that navigators passing through the Strait of Magellan had used freely, finding it quite good.

Among the composites, he found a cudweed with large cottony leaves, naming it *Gnaphalium pyramidale* Thou. He called a small everlasting in the turf *Xeranthemum caespitosum* Thou. [=*Chevreulia stolonifera* Cass.]. At high elevation on the mountain he climbed, he found one of the smallest plants among the composites that he named *Calendula pusilla* Thou. [=*Lagenifera commersonii* Cass.

An abundant, low, climbing shrub, with leaves like those of mountain-ash or silver-weed, he named *Ancistrum sarmentosum* Thou. [=*Acaena sanguisorbae* Vahl]. He also believed he had observed a new genus in the Rubiaceae, close to *Nertera* or *Gomozia*, and which encompassed two species. They are creeping herbaceous plants with inconspicuous, solitary, flowers terminating the branches; the fruit was a small berry, brilliant red, unpalatable. His proposed new genus, *Erythrodanum*, did not stand; and both his new species were referred to *Nertera depressa* Banks & Soland. ex Gaertn. He fared better with *Hydrocotyle capitata* Thou. and with his sub-woody *Rumex frutescens* Thou.[112]

As for monocotyledons, he found only a few, all members of Cyperaceae and Poaceae. Among the four sedges, he left his two carices unnamed; and the other two are now placed in *Uncinia brevicaulis* Thou. He reported two species of bulrush. A small one, *Scirpus prolifer* Thou. [=*Scirpus thouarsianus Schult.*], was widespread, forming very thick turf, and arid slopes had no other cover than tight, distinct mats of that species. The second bulrush he named *Scirpus sulcatus* Thou.

A turf was also formed by a very small canary grass, *Phalaris caespitosa* Thou. A rarer plant *Phalaris mollis* Thou. [=*Polygonum intermedius* Carmich.], rose to a height of 2'. The plant had a very soft texture, indicating it would provide a very good forage. Among all the grasses, the most remarkable for size and abundant growth was within a genus characterized by a single style, *Ponceletia arundinaceae* Thou. [=*Spartina arundinaceae* (Thou.) Carmich.]. It had the habit of a sand rush and grew 4-6' high. So abundant was it on fertile terrain that it nearly blocked one's passage. It also grew on the highest summits where it gave him the greatest assistance. Because of its strong roots, it helped him to clamber up the slopes.[113]

Cryptogams were very imperfectly understood in the 18th century, and it is hardly surprising that he could do little of value with mosses, lichens, and fungi. Although he did attach names to some of those he found, the specific names had already been used by other authors. Bryologists and lichenologists appear to have

112. Ibid. pp. 15-17, 38-46.
113. Ibid. pp. 17, 35-37.

ignored them with good reason. But, as he was much more comfortable with vascular plants, his list of ferns on Tristan da Cunha did attract botanical attention. Most required transfer, but a few of his specific names have survived:

no. 61 *Pteris vittarioides* Thou., fig. no. 1 [=*Vittaria vittarioides* (Thou.) C. Chr.

no. 62 *Acrostichum succisaefolium* Thou. [=*Elaphoglossum succisaefolium* (Thou.) Moore

no. 63 *Acrostichum laurifolium* Thou. [=*Elaphoglossum laurifolium* (Thou.) Moore

no. 66 *Polypodium aquilinum* Thou. [=*Dryopteris aquilina* (Thou.) C. Chr.

no. 67 *Polypodium tomentosum* Thou., fig. no. 3 [=*Dryopteris tomentosa* (Thou.) O. Ktze.

no. 75 *Tricomanes aeruginosum* Thou. [=*Hymenolepis aeruginosum* (Thou.) Carm.[114]

Such were the plants he found naturally established in the soil of Tristan da Cunha. He suspected that many of them could have become naturalized as the island was placed between Africa and South America, and as the vegetation adhered to both of those climates. The island had *Phylica* and *Pelargonium* in common with the Cape of Good Hope, while the ferns and the *Ancistrum* [=*Acaena*] seemed related to South America. He had been surprised to find one specimen of what appeared to be *Chenopodium album* L., which he would have recognized as European, but could find only the one individual.

There were other signs of earlier visitors to the island. To his great pleasure, he found a small piece of ground that had been cleared in the past. All that remained from whatever had been planted were some lettuce and turnip plants that were both in flower and in seed. He identified them as *Lactuca sativa* L. and *Raphanus sativus* L. During the subsequent passage to the Cape of Good Hope, he learned they could have dated from an American ship known to have stopped at the island for fishing, probably for whaling.[115]

Dupetit-Thouars reached the Cape of Good Hope at the beginning of February 1793, remaining about two weeks before proceeding to Mauritius. In none of his subsequent publications does he tell us where he learned of his brother's fate; but as there can have been no record of Aristide anchoring at Cape Town, one would suppose he had to expect a disaster at sea. In any case, he went on to

114. Ibid. pp. 30-34.
115. Ibid. pp. 18-19.

Mauritius where he would at least be among compatriots, and where he would have to recognize that the search for La Pérouse had aborted.

He arrived, of course, without his baggage and virtually penniless. He would remain on the African islands from 1793 to 1802, leaving no detailed account of his botanical exploration. It is said that he was funded for two years by several wealthy planters on Mauritius, thereafter obtaining a minor official position that provided subsistence but no salary. That arrangement enabled him subsequently to spend 6 months on Madagascar and, later, 3 1/2 years on Reunion. These ventures were followed by interims of high fever lasting for months. His ultimate return to France was delayed by war between Britain and France. With the reestablishment of peace in 1802, the government granted him passage on a frigate bound for France. He landed at Rochefort in early September with his herbarium of about 2000 specimens and about 600 drawings of those plants that had most interested him.[116]

The arrival was timely. After Aristide's release from Portuguese custody, family members had petitioned the Portuguese government to recognize its obligation to provide an indemnity to the ruined family. Not until 1802 did a sister's persistent efforts prevail. The award for damages came to 334,000 francs. A notice of the award was put in the press to notify subscribers they could apply to recover losses. Dupetit-Thouars, at least, was no longer penniless, but he faced reduced circumstances compared to his earlier life.[117] His applications for government funds to publish the material he had brought back were fruitless. That may at least partially account for his decision to limit his publication to new genera from the African islands, substantially reducing the cost of publication.

Any botanist who means to review Dupetit-Thouars' new genera from Madagascar may be disconcerted by his habit of republishing some of them in successive papers. In general, moreover, he published a new genus without indication of the type species. In the lists below, only the genera that still stand are included, and only in the order they were initially published. The new genera were fully and impressively diagnosed, suggesting a discriminating eye. He acknowledged having had access to the Commerson herbarium in Paris for comparative material. A prior collector of several specimens, Father Francisco Noronha, a Spanish physician and botanist, who had visited Madagascar, he identified as Norôna.

116. Boreau. 1845. "Notes sur la vie et les travaux du botaniste Aubert du Petit-Thouars," pp. 59-61.

117. Aubert du Petit-Thouars. 1811. "Description abrégée des isles de Tristan d'Acugna," p. 34; Richard. 1986. *Le Voyage d'Entrecasteaux*, p. 27.

Histoire des végétaux recueillis sur les îles de France, La Réunion (Bourbon) et Madagascar.
Paris: Mme Huzard, 1804

p. 23. *Didymeles* Thou. in Didymelaceae
p. 25. *Ptelidium* Thou. in Celastraceae
p. 31. *Dicorypha* Thou. In Hamamelidaceae
p. 33. *Bonamia* Thou. in Convolvulaceae
p. 35. *Monimia* Thou. in Monimiaceae

Histoire des végétaux recueillis dans les îles australes d'Afrique.
Paris: Levrault, Schoell, 1805

p. 37. *Sarcolaena* Thou. in Sarcolaenaceae
p. 41. *Leptolaena* Thou. in Sarcolaenaceae
p. 43. *Schizolaena* Thou. in Sarcolaenaceae
p. 47. *Rhodolaena* Thou. in Sarcolaenaceae
p. 49. *Macarisia* Thou. in Rhizophoraceae
p. 51. *Asteropeia* Thou. in Theaceae
p. 59. *Paropsia* Norôna ex Thou. in Flacourtiaceae
p. 61. *Deidamia* Nor. ex Thou. in Passifloraceae

Histoire des végétaux recueillis dans les îsles australes d'Afrique.
Paris: Tourneisen fils, 1806. [Republishes everything in 1804-1805 editions, but with far superior plates.]

p. 68. *Amelosycios* Thou. in Cucurbitaceae. Fig. xxii shows *A. scandens.*
p. 69. *Zymum* Nor. ex Thou. [=*Tristellateia* Thou.] in Malpighiaceae

Genera Nova Madagascariensis, secundum Methodum Jussiaeanum Disposita.
Paris: n.p., 1806

p. 2. *Dicraeia* Thou. in Podostemaceae
Hydrostachys Thou. in Hydrostachyaceae
Halophilia Thou. in Hydrocharitaceae
p. 3. *Tristicha* Thou. in Podostemaceae
p. 5. *Potameia* Thou. in Lauraceae
p. 6. *Barbeuia* Thou. in Barbeuiaceae
Physena Nor. ex Thou. in Physenaceae
p. 7. *Dilobeia* Thou. in Proteaceae

p. 10. *Voaganga* Thou. in Apocynaceae
p. 11. *Alafia* Thou. in Apocynaceae
Stephanotis Thou. in Asclepiadaceae
Plectaneia Thou. in Apocynaceae
p. 12. *Aspilia* Thou. in Compositae
p. 14. *Tristellateia* Thou. in Malpighiaceae
p. 18. *Burasia* Thou. in Menispermaceae
p. 19. *Spirospermum* Thou. in Menispermaceae
p. 20. *Brexia* Nor. ex Thou. in Grossulariaceae
p. 22. *Trilepisium* Thou. in Thymelaeaceae
Intsia Thou. Leguminosae
p. 23. *Voandzeia* Thou. in Leguminosae
Dichapetalum Thou. in Dichapetalaceae
Sorindeia Thou. in Anacardiaceae
p. 26. *Macaranga* Thou. in Euphorbiaceae

None of the above new genera were in the Orchidaceae despite Dupetit-Thouars' special interest in orchids. Several years later, he read a paper before the Institut in which he proposed an alteration in the classification of orchids, as previously used by Linnaeus and successors, that would accommodate better the many new plants he had studied on the African Islands. He further proposed an alternative nomenclature in which each genus would terminate with the suffix *orkis*, not *orchis*. Finally, he proposed a new genus, *Corymborkis* Thou., which continues to stand with his unique spelling, and now a genus of 20 species.[118]

More than a decade later, he returned to the topic in a short treatise. In his dichotomous keys, he showed his division of orchids into 3 sections, initially based upon 3 variations in the structure of anthers. The keys offered alternative names for each genus, the traditional and his own; but he abandoned the *k* for *ch*. Thus, his own new genus appeared as *Corymborchis corymbosa* Thou.[119]

Dupetit-Thouars' situation had unexpectedly altered in 1807 when he was nominated to fill the vacated directorship of the imperial nursery of La Roule in Paris (within what is now a *quartier* in the Eighth Arrondissement). The position

118. Aubert du Petit-Thouars. 1809. "Extrait de trois mémoires lus à la première classe de l'Institut, sur l'histoire des plantes Orchidées des îsles australes d'Afrique." *Nouveau Bulletin des Sciences*, par la Société Philomatique 1, no. 19 (April): 314-319.

119. Aubert du Petit-Thouars. 1822. *Histoire particulière des plantes orchidées recueillis sur les trois îles australes d'Afrique; de France, de Bourbon, et de Madagascar*. Paris: Chez l'auteur, unpaged. See Ib Friis & Finn N. Rasmussen. 1975. "The Two alternative systems of nomenclature proposed and used for orchids by Du Petit-Thouars, with special regard to *Coymborkis* Thouars." *Taxon* 24 (May): 307-318.

provided him an annual salary of 3,600 francs and ended his financial difficulties. Founded in 1741, the nursery in time became the primary center for acclimating exotic trees for the reforestation of France. But after Louis XVI purchased the domain of Rambouillet to make it the primary center for agricultural experimentation, Le Roule had received only the surplus plants and seeds sent by André Michaux from America.

The ease with which Dupetit-Thouars dedicated himself for the next 20 years to agricultural and horticultural matters seems to confirm his claim that the quest for new foods for the improvement of French agriculture had been of interest to him during his 10 years of botanizing abroad. The new appointment was a congenial opportunity to work for that advancement through experimentation and the delivery of public lectures. Such lectures mainly featured horticultural topics; his experiments were largely in plant physiology, providing material for papers read before learned societies.[120]

He espoused a theory of plant morphology in 1828 based upon the homology between the cotyledons, the foliage leaves, and the floral parts of a plant. The theory was quite similar to that proposed by Goethe in *Die Metamorphose des Pflanzen* in 1790, which may have been derived from Linnaeus's cortex-medulla theory or from C.F. Wolff and J.F. Turpin. A few botanists had similar ideas, including Robert Brown, so that it is unknown whether such ideas were acquired independently or directly from Goethe. The point is not that an actual leaf becomes modified into a floral part, but that there is a common initial process within the plant that is modified to produce different ends. The implication was that Nature can produce varied forms through modification of a single organ, what Goethe meant by *metamorphosis*. This homology was advocated in France by Dupetit-Thouars before botanists in general were interested.[121]

He complained that he had experienced a "coldness" in botanical circles, attributing it to his advocacy of unorthodox theories. The complaint overlooked the fact that he had been named a chevalier in the Légion d'Honneur and had been elected to a seat in the Académie des Sciences on 10 april 1820, measures of considerable esteem. Even so, his colleagues could scarcely have avoided regarding him as enigmatic and unsociable.

120. Aubert du Petit-Thouars. 1825. *Notice historique sur la pépinière du Roi au Roule*. Paris: L'Imprimerie du Gueffier, pp. 6-10, 25.

121. Boreau. 1845. "Notes sur la vie et les travaux du botaniste Aubert du Petit-Thouars," pp. 64-65.

His mania for plants was evidently extreme, to a point where he often did not recognize friends or even members of his family. His preoccupation with his work came to deprive him of any normal sense of order, and visitors to his cabinet found it to be a true chaos. More than once he went out in dressing gown and without a hat to attend sessions of the Académie des sciences. We already have noted that it was not unusual to find his materials published more than once, and he left numerous uncompleted manuscripts, suggesting an inability to pursue lengthy projects to their end. Yet, those he did publish reveal him to have been an acute observer.[122]

The unexpected decision of the royal government in 1827 to terminate the subsidies to the state nurseries came as a profound shock to him, and he soon died, in Paris, 12 May 1831.

122. Ibid., pp. 61-64.

Portrait of Louis-Marie-Aubert Aubert Dupetit-Thouars.
© 2000, Hunt Institute, CMU, Pittsburgh, PA 15213. U.S.A.

VII
The Voyage of Entrecasteaux: Part 1
Genesis & Preparations

La Pérouse had sailed from Botany Bay on 10 March 1788 to be heard from no more. The first public expression of concern was made in a paper read to the Académie des sciences by Alexandre de Laborde in 1790. Early in 1791, a petition was prepared by the members of the Société d'Histoire Naturelle for the rescue of La Pérouse for presentation to the National Assembly. André Thouin, who was treasurer of the Société d'Histoire Naturelle, and who had taken a personal part in the preparation of the La Pérouse expedition, was the principal instigator of the petition. He was also pushing for a continuation of geographic and scientific research.[123]

The tenor of both the petition, and the legislative response to it, remain useful revelations of the spirit of the Enlightenment. "For two years," the petition began, "France has awaited the return of Monsieur de La Pérouse in vain. Those who are concerned for his person and about his discoveries have no knowledge of his fate. Alas! The fate that they suspect is perhaps quite more frightful than the one he has met. Perhaps he has escaped death only to be subjected to the continual torments of a revival and a shattering of hope. Perhaps he has run aground on some island in the South Seas from where he holds out his arms toward his country, waiting in vain for his liberator.

"It was not for trivial ends, nor for personal gain, that Monsieur de La Pérouse exposed himself to so many perils. And the decent nation that expected to reap the benefits of his labors also owes him its concern and its assistance.

"We learned earlier of the loss of a number of his companions, either swallowed by the seas or murdered by aborigines. Support our remaining hope of rescuing those of our brothers who have escaped the fury of the waves or the frenzy of cannibals, so that they may return to our shores to be overcome with joy when touching this free land."[124]

The preamble to the Decree of 9 February 1791 acknowledged that there was no longer any reason to hope that La Pérouse's ships still ploughed the surface of

123. Letouzey. 1989. *Le Jardin des Plantes*, p. 227.
124. Milet-Mureau. 1797. *Le Voyage de La Pérouse* 1: liii-liv.

the seas. The navigator and his companions must be stranded on some desolate beach, perhaps threatened by climate, animals, other human beings, or Nature, appealing to their country that alone can envision their misfortune. "If you prefer such an idea to merely accepting the idea of their loss, and are sensitive to the appeals made by M. de Laborde and the naturalists of Paris, then we must fly to the aid of our brothers: Humanity requires it!"

The preamble made no effort to conceal the difficulties in searching such a vast area, but found no excuse in that fact for not attempting the search. "We owe such a concern to men who have dedicated themselves; we owe it to the sciences which await the benefits of their research. That concern is increased by the fact that M. de La Pérouse was not one of those adventurers who instigate great enterprises in the hope of making a great name for themselves or in the expectation of securing private wealth. He did not seek the command, would have preferred to refuse it, and his friends knew that he was merely resigned to it." As for the difficulties of the search, La Pérouse did transmit his proposed route, and that should assist whoever should be entrusted with the search.

"Such actions bring honor," the preamble concluded, "to the nation that recognizes they must be undertaken; and the humanitarian sentiment that gives rise to them will characterize our century. No longer does the European penetrate the most distant latitudes to invade or to devastate, but rather to bring benefits and dividends; no longer the grasp for corrupting metals, but rather for the conquest of those useful plants that can render man's life easier and happier."[125]

The Decree of 9 February 1791: "The National Assembly, after having heard its assembled committees on agriculture, commerce, and the navy, decrees that the king shall be urged to issue orders to all ambassadors, residents, consuls, and agents of the nation accredited to the various powers authorizing them in the name of humanity, the arts, and the sciences, to engage the various sovereigns to whom they are accredited to order all navigators and other agents within their jurisdictions, and wherever they may be, but especially in the southern part of the South Seas, to make every effort to find the two French frigates, the *Boussole* and the *Astrolabe*, commanded by M. de La Pérouse, along with the ships' companies, and to pursue any inquiry that could establish either their survival or their shipwreck. So that, in the event M. de La Pérouse and his companions should be found or encountered, no matter where, they could be given every assistance and provided with all the means to return to their native land, as well as enabling them to bring back everything then in their possession. The National Assembly here pledges to indemnify and even to recompense, according to the importance of the service

125. Ibid. 1: lv-lx.

rendered, whoever shall give help to these navigators, procure news of them, or simply arranges a restitution to France of whatever papers and effects that might have belonged, or did belong, to their expedition.

"Decrees furthermore that the king be urged to have one or several ships outfitted on which scientists, naturalists, and illustrators will be embarked, giving the commanders of the expedition the double mission of searching for M. de La Pérouse, in accordance with documents, instructions, and orders that will be given to them; while, at the same time, undertaking research relating to the sciences and to commerce, by taking every measure, quite apart from the search for M. de La Pérouse, or even after having found him or obtaining news about him, that will render this expedition useful and advantageous to navigation, to geography, to commerce, to the arts, and to the sciences."[126]

In Chapter 2 the reader was given indication that the royal instructions for Entrecasteaux would be identical to those earlier prepared by Claret de Fleurieu for La Pérouse. All Fleurieu had to rewrite for Entrecasteaux, consequently, was a new itinerary, or Navigational Plan, that was informed by the route La Pérouse had projected as he was preparing to sail from Botany Bay. He would backtrack to the Tonga Islands, going by way of the southern part of New Caledonia; sail northward to Santa Cruz Island, the Solomon Islands and the Louisiade Archipelago; pass between New Guinea and Australia to explore the Gulf of Carpentaria; then down the west coast of Australia and along the south coast as far as Tasmania. He would turn back in order to reach Mauritius by December of 1788.

If neither La Pérouse nor any traces of his expedition had been found after following that route, and the primary goal of the voyage had to be abandoned, "at least the sieur d'Entrecasteaux will enjoy the satisfaction and the glory of having contributed infinitely to the perfecting of geography and to the growth of our knowledge about mankind. By making navigation on seas, very little frequented up to now, more secure through more reliable information, he will have opened new channels to the commerce of the European nations and will have acquired an eternal right to their gratitude. . . . It must be understood from the outset that, while the prescribed plan is desirable, His Majesty recognizes that unforeseen contingencies will arise that make alterations advisable. Given Entrecasteaux's great experience on the open seas, he is authorized to make adjustments if he believes they will not essentially harm the principal goal of his mission nor extend the length of his voyage beyond the prescribed three years."[127]

126. Elisabeth-Paul-Edouard, chevalier de Rossel, ed. 1808. *Voyage d'Entrecasteaux, envoyé à la recherche de La Pérouse*, 2 vols. Paris: Imprimerie impériale. 1: xxvii-xviii. Also see Stafleu. 1967. *Adanson, Labillardière, De Candolle*, pp. 21-22.

127. Ibid. 1: xxi-xliv.

By choosing Entrecasteaux to command the rescue and scientific mission, Fleurieu again demonstrated both his acute judgment of men and his unquestionable commitment to the scientific and humanitarian aspirations foundational to both expeditions. Such commands required special qualities of mind and morality; they could not be subject to routine naval advancement based upon seniority or social prominence. Antoine-Raymond-Joseph de Bruny d'Entrecasteaux (1737-1793) was, in fact, a descendant of a prominent Provençal noble family; but his naval service had testified to both unusual capacity and honor.

His family, originally Bruni, had moved into southern France in the 17th century, becoming barons of La Tour d'Aigues, north of Aix-en-Provence. Our naval captain belonged to the cadet branch of the family that had purchased the marquisat d'Entrecasteaux in 1714 for 250,000 livres. Thereafter, the family was easily identified with Aix as his father, brother, and nephew all served consecutively as *président à mortier* of the Parlement de Provence (i.e., Jean-Baptiste 1701-1792, Jean-Paul 1730-1794, and Jean-Baptiste-Raymond-Joseph-Guillaume 1758-1785.)[128]

Consequently, our expedition commander is typically cited as having Aix as his birthplace when it actually was the château d'Entrecasteaux, about 75k to the east. He received his education from Jesuits and would retain his Catholicism all his life. But, as was fitting in his era, he was opposed to religious intolerance. As a second son, he was admitted to the navy at age 16 in 1754 as garde de la marine, a midshipman. He saw wartime naval service in the Indian Ocean during the Seven Years' War and the American Revolution, thereafter serving in peacetime as deputy director of Ports and Arsenals under Fleurieu.[129] He had by then acquired the reputation for being the very model of virtue, integrity, and propriety.

At that juncture, he received the dismaying news that his nephew in Aix had murdered his wife on the night of 30-31 May 1784. The nephew had then fled the country to avoid the legal consequences, somehow reaching Portugal. He would die the following year in a Lisbon prison, 16 June 1785, for reasons obscure.[130] The scandal had magnified, meanwhile, because of suits filed by the murdered woman's father, Jean-Baptiste de Castellane, marquis de Grimaud, charging the murderer's father, Jean-Paul de Bruny, with complicity in the crime and the escape, plus seeking custody of the two surviving daughters.

128. Jean Audouard. 1912. *Une Famille provençale au XVIIIè siècle, les Bruny d'Entreasteaux*. Paris: H. Daragon, p. 177.
129. Etienne-Gabriel-Joseph, baron Hulot. 1894. *D'Entrecasteaux, 1737-1793*. Paris: Société de géographie, pp. 1-5; Horner. 1995. *Looking for La Pérouse*, p. 15.
130. Audouard, 1912. *Une Famille provençale*, p. 126.

Captain d'Entrecasteaux was stunned to find his name tarnished by such an abominable crime and requested immediate retirement from the service. Maréchal de Castris refused the request, saying that the country could not be deprived of the services of a man of his merit. Entrecasteaux then asked for an assignment overseas to get away from France despite deep attachment to his father and brother. Consequently, he was given command of the Indian Station, which meant that he also became governor of Mauritius.

While he was grateful for the appointment, before long he began receiving letters from home that unceasingly pleaded for his return. In time he began to reproach himself for having abandoned his relatives to confront their shame and difficulties alone, which culminated with a letter requesting recall (8 November 1788). Despite his return, the family situation in Aix became intolerable. His elderly father, Jean-Baptiste de Bruny, retired to his marquisat of Entrecasteaux (where he would die in February, 1792). His brother, Jean-Paul de Bruny, even though ultimately acquitted of the charges of complicity that had been brought against him, believed he must leave his native Aix permanently. He retired to Carpentras in 1789. (He would be arrested in nearby Avignon, 29 November 1793, and executed as a counterrevolutionary in Orange on 20 June 1794.)[131]

Captain d'Entrecasteaux's official letter of appointment to command the rescue mission came from the new minister of the marine, Antoine-Jean-Marie Thévenard, 28 May 1791, but Fleurieu provided continuity as director of both expeditions. The king, continuing to take a personal interest in the voyages, authorized Fleurieu to promote Entrecasteaux to the rank of rear-admiral before his departure; and Entrecasteaux was to have the right to propose the names of those officers and midshipmen he felt appropriate to serve him.[132]

The two ships assigned to him for the voyage, the *Recherche* and the *Espérance*, although called frigates, had really been flutes designed as naval cargo carriers. They had been refitted as if they were frigates, but they remained slow ships unlike genuine frigates, and Entrecasteaux would later have good reason to complain about them as being "bad sailors." He would command the expedition from the *Recherche*. He nominated Alexandre d'Hesmivy d'Auribeau to be captain of the *Recherche*, Pierre Renard to be chief-surgeon; and Louis Ventenat as chaplain, who could double as a naturalist. For the *Espérance*, he chose Jean-Michel Huon, chevalier de Kermadec, as captain; Denis Joanet as chief-surgeon; and the Benedictine Ambrose Pierson to be chaplain, who could double as an astronomer.

131. Jean Audouard. 1910. *Un Drame passionnel à la fin du XVIIIè siècle: le crime du marquis d'Entrecasteaux.* Paris: H. Daragon, pp. 79-81.
132. Hulot. 1894. *D'Entrecasteaux*, pp. 84-94.

The scientific personnel, chosen for him as in the case of the voyage of La Pérouse, was not subject to naval regulations, a potential source for friction. As La Pérouse had learned, it was fruitless to appeal to the academies to take into account a congenial personality when nominating savants for employment on board ships. Indeed, that quality may have been in short supply. Among the seven members of the scientific staff on the *Recherche*, two were botanists: Jacques-Julien Houton de Labillardière, who would maintain and publish a valuable journal, and Louis-Auguste Deschamps; a gardener, Félix Delahaye; and a now obscure painter named Piron. Among the five on board the *Espérance*, Claude Riche was the naturalist and the obscure Chailly-Ely the painter. That Labillardière, Deschamps, and Riche were all doctors of medicine provided additional expertise for the observation of aboriginal populations by the naval surgeons.[133]

La Pérouse, to be sure, had reported unofficially to the navy, before leaving Botany Bay, about his difficulties with his scientific personnel: a portrait of probable friction between an aristocracy of learning and an aristocracy of birth, quite antecedent to the events of 1789. For those who had seen themselves as the representatives of the Enlightenment, sympathetic to a substantial reform of both society and the regime, the actual revolutionary changes by the autumn of 1791 could broaden the terrain for hostilities and exacerbate the passions. Authors who have commented previously about such hostilities, which would become a prominent feature of the Entrecasteaux expedition, have essentially converted the aristocrats of learning (by 1791) into republicans and the aristocrats by birth into royalists: an easy conversion made by those historians who are convinced that allegiance or belief are determined by membership in a group. In this particular case, the even easier assumption has been that the republicans had all been liberal, the royalists all reactionaries.

A look at individual cases on board the *Recherche* and the *Espérance* will, indeed, provide numerous instances where an individual conforms to his stereotype. But a sufficient number of deviations from the rule raises legitimate doubt about its reliability. A factor further confusing the genesis of loyalties derives from the fact that the naval officers on board, some aristocrats and some commoners, had all taken an oath to serve the king; and the degree to which that remained a binding factor after 1789 resists generalization.

What is more, regardless of social origin, they had all observed, as commissioned officers, widespread attempts around Brest by local radicals to encourage naval crewmen to disobey their officers' orders in the name of the Revolution.

133. Jacques-Julien Houtou de Labillardière. 1800. *Relation du voyage à la recherche de La Pérouse*, 2 vols. Paris: H. J. Jansen. 1: vi-xiv; Horner. 1995. *Looking for La Pérouse*, pp. 37-41.

Their outrage did not vanish once they were at sea, and they remained sensitive to the possibility that the necessary discipline to maintain order at sea was vulnerable to disruption by anyone suspected of republican sympathies.

The case of the commander-in-chief himself is illustrative: an aristocrat by birth, but certainly a man of liberal outlook, quite evidently imbued with the humanitarian and tolerant spirit of the Enlightenment, yet retaining a strong sense of loyalty to his king. His anticolonialism, a legacy of Western European liberalism and earlier manifested in La Pérouse's pronouncements about Spanish rule in Concepcion, Monterey, and Manila, would emerge in Entrecasteaux's critical observations on conditions resulting from the governance by the Dutch East India Company.[134]

If we turn to the three botanists on the scientific staff, two of them would be soon identified as republican, Labillardière and Riche; whereas Deschamps apparently remained a royalist. Their painter, the draughtsman Piron, now so obscure that not even his first name is known, became identified as republican at the time. Certainly his drawings, used by Labillardière in the publication of his *Relation*, reflected neither luxury nor sensuousness, but rather the republican virtue of heroic stoicism made fashionable by Jacques-Louis David: a neoclassical perfectionism that was certainly no reflection of aboriginal existence.[135]

Two of the remaining naturalists had enlisted as chaplains and, therefore, were officially naval personnel and subject to naval regulations. Pierson, an astronomer, had taken the oath to uphold the Civil Constitution of the Clergy required by the revolutionary government. Using generic classification, such an action could have been labeled as an outward sign of republican sympathies. It also happens to have been a frequent indication of clerical opportunism. We do not know in Pierson's case.

Ventenat, the second chaplain and a botanist, owed his appointment to the influence of his older brother, Etienne-Pierre, a man prominent in Parisian botanical circles, as a compensation for the loss of his benefice in Limoges following his *refusal* to take the oath to uphold the Civil Constitution of the Clergy. His refusal could easily have been interpreted to be a reflection of royalist sympathy. In fact, his motives remain unknown. On board the *Recherche*, he would early win the reputation for being a conniving trouble-maker and came to be regarded as a republican by the royalists, whatever he actually may have been. On one point we may be reasonably certain. The issue over whether or not to take the

134. Rossel. 1808. *Voyage d'Entrecasteaux*. 1: 155-166.

135. Smith. 1985. *European Vision and the South Pacific*, p. 149; Roger Collins. 1989. "Voir avec d'Entrecasteaux: l'oeuvre du dessinateur Piron." Ulane Bonnel, ed. *Fleurieu et la marine de son temps*. Paris: Economica, p. 55.

oath deeply and permanently divided the French clergy into hostile camps, and our two chaplains cannot have been congenial colleagues.[136]

Once Labillardière, a friend of André Thouin, had been designated to be a botanist on the voyage, Thouin nominated an experienced gardener, Félix Delahaye, to be the gardener for the expedition. He had been born in 1767 into a farming family in Caumont, a village near Le Havre, and had left home at 17 to work in the Jardin botanique de Rouen. About five years later, 1789, he had moved on to the Jardin du roi in Paris to be a student of André Thouin. Consequently, Delahaye knew not only gardening techniques but was well acquainted with the plants culivated at the Jardin du roi; and he had shown a disposition to travel.[137]

As it is known that the royal instructions for scientific exploration, originally prepared for La Pérouse, were used except for the itinerary for Entrecasteaux, it may be assumed that Thouin's botanical and horticultural recommendations, prepared for Fleurieu, were also transmitted to Entrecasteaux. Some confusion on the matter could arise from Letouzey's claim that Thouin's instructions reached Brest days *after* the expedition had sailed. That would imply that Delahaye embarked without horticultural equipment or seeds, untenable in view of his considerable activity during the voyage.

Letouzey also noted, fortunately, a letter written by Delahaye to Thouin from Brest dated 27 September 1791, the day before the expedition sailed, in which he commented that he had not received Thouin's *final instructions*. Last-minute additions, in other words, not the principal instructions for stocking the ship with necessary supplies and equipment. Delahaye took the opportunity to complain that he had found himself the only naturalist who would not be eating at the headquarters table, where the food would be better than at the masters' table to which he had been assigned.

The distinction would not have surprised Thouin. He had worked closely with Antoine-Laurent de Jussieu at the Jardin du roi since the replanting of the demonstration garden in 1774. During the ensuing years, they both would hold professorships at the Jardin des plantes and be colleagues in the Académie des sciences without Jussieu ever using familiar address with Thouin. The distinction between savant and gardener was recognized from beginning to end; and Thouin, who wanted nothing more than the freedom to cultivate his garden, gave no sign of resentment and remained a happy man in that freedom. We are reminded that the spirit of the Enlightenment was liberal, not egalitarian.[138]

136. Horner. 1995. *Looking for La Pérouse*, pp. 37-41.

137. Auguste Chevalier. 1953a. "Félix Delahaye, le jardinier de l'expédition envoyée à la recherche de La Pérouse." *Revue internationale de botanique appliqueée et d'agriculture tropicale*. 32: 63.

138. Letouzey. 1989. *Le Jardin des Plantes à la croisée des chemins avec André Thouin*, pp. 227-238.

It now seems likely that Thouin's final instructions, presumably dated 3 September 1791, were sent to Fleurieu as procedure required; and that Fleurieu recast them in a letter to Entrecasteaux, dated 13 September 1791, which embodied views about the efficient management of both naturalists and gardeners on such expeditions held by both Thouin and Fleurieu. Namely, how to utilize their specialties and to avoid overlapping employments.

"The gardener," the letter began, "is not counted as one of the naturalists. His duties are to sow seeds from Europe in the lands where you go ashore where it appears they will prosper, and to indicate to the natives of the country, so far as he can, the manner of cultivating and multiplying them. Those productions that can provide subsistence for man must command his attention in particular. The culture of plants or useful shrubs that it will be possible to transport alive to our climate must be confided especially to his care; as well as procuring, as a great service to our colonies, breadfruit and other nutritive productions native to the islands in the open seas. You should leave them on Mauritius where they can be cultivated and developed for transplanting on our islands in America. Beyond which the gardener should vigorously assist the naturalists' research; but such duties are necessarily subordinate to theirs, as their profound study is what makes an appreciation of the usefulness of each natural production possible."

Fleurieu added that the naturalists had the duty to oversee the drawing of all objects pertaining to natural history: plants, trees, and animals. He (or was it Thouin?) believed that the illustrators themselves would soon sense the benefits to be achieved when the distinctive and characteristic parts of each object to be drawn were made clear to them. "And all of them will forget personal advantage or achievement when working together for the glory of the nation and for the improvement of human knowledge."[139]

Among the naturalists employed for the Entrecasteaux expedition, only Jacques-Julien Houtou de Labillardière would perform according to optimistic expectations. He would make an immense collection of plants, would be intensely interested in the native peoples, and would publish a record of his voyage as well as his new genera and species. Only the gardener, Delahaye, would match him in dedication. But it is only fair to add that Labillardière alone had had extensive prior experience in the field as an incomparable preparation for the rigors of the voyage. The botanical credentials of the remainder of the team were sufficiently minimal as to permit wonder about the reasons behind their appointments, which remain obscure. As a consequence, to record the botanical history of the expedi-

139. Rossel. 1808. *Voyage d'Entrecasteaux.* 1:xliv-xlv.

tion, one is virtually limited to the journals kept by Labillardière and Entrecasteaux.

Baron Hulot, a member of the Société de géographie, who undertook his study of the expedition on behalf of that society as part of the centennial of Entrecasteaux's death, openly preferred Entrecasteaux's account to Labillardière's. He did not mean to denigrate the value of Labillardière's scientific work, which he recognized to have been admirable. He simply found the tone of his narrative to be partial and vindictive in contrast to that of Entrecasteaux, described as having been "uniformly liberal and tolerant."[140]

Fair enough as far as Entrecasteaux was concerned. But tone is not necessarily a valid measure of reliability. There had been, in truth, initial doubts about the wisdom of Labillardière's appointment as he had been well known for a caustic wit, easily construed as the sign of a vindictive spirit, but more often indication of a quick mind married to impassioned probity. The alternative to taking the highly competent, but potentially difficult, person on board was to recruit another botanist like Louis-Auguste Deschamps. So little did Deschamps accomplish that it is even unknown whether he bothered to make a herbarium. Ventenat's observation that Deschamps had never let field work interfere with his three meals a day on shipboard, while others struggled with the dense foliage and swamps to do their collecting, has a ring of truth.[141]

In short, the political and religious tensions of that moment, if troublesome enough, were not solely to blame for the antagonisms on that voyage, which Entrecasteaux to his credit endeavored to constrain, and which Labillardière more bluntly exposed in the candor of his narrative. Some might describe the narratives as reflections of a naval-royalist and a republican-scientist. Others may be struck by the credibility of two men of evident integrity, whose narratives are equally entitled to critical examination.

Jacques-Julien Houtou de La Billardière was born in Alençon, now Orne, on 25 October 1755. The family was provincial bourgeois, seigneurial to be precise, and some ancestor had apparently added the La Billardière to the name, presumably a property he owned in Normandy. Our subject seems to have dropped the family name, Houtou, about 1780, keeping La Billardière as his name, later simply Labillardière. His gentle origin gave him a private income for life, enabling him to devote his entire life to science without having to become a professor. Little reliable information survives about his early years except that he became a student in the

140. Hulot. 1894. *D'Entrecasteaux*, p. 113.

141. Brian Plomley & Josiane Piard-Bernier. 1995. *The General: the Visits of the Expedition led by Bruny d'Entrecasteaux to Tasmanian Waters in 1792 and 1793*. Launceston: Queen Victoria Museum, pp. 23, 345-354.

Collège royal d'Alençon, a Jesuit institution until 1763. Thereafter, the teaching would have been done by secular clergy. To judge only by his later excellent use of Greek and Latin in botanical diagnoses, he must have benefited from a good classical education.

After the conclusion of those studies, about 1772, Labillardière went to Montpellier to take his medical degree, where he became the devoted student of Antoine Gouan. The Linnaean sexual system, advocated by Gouan, would serve Labillardière for the remainder of his botanical career. We next locate him about 1780 in Paris taking additional study in medicine and where he came to be a friend of the Breton René Louiche Desfontaines. It would be the most durable friendship of Labillardière's life and was almost immediately beneficial.

The reader may recall that Desfontaines, five years older than Labillardière, had become eminent very young with the assistance of the king's physician, Dr. Lemonnier. Thanks to Desfontaines, Labillardière was soon brought to the attention of Lemonnier, L'Héritier de Brutelle, and Jussieu, all of them sources of patronage and encouragement. Without much doubt, Lemonnier made it possible for Labillardière to spend two years, 1782-1784, in England for the study of exotic plants already cultivated there but still unknown in France. He also made the acquaintance of Joseph Banks, an association that would serve him well in later years.

After leaving London, Labillardière headed for the French Alps, obtaining the guidance of Dominique Villars for traveling through Dauphiné. Moving on into Savoy, he put himself under the guidance of C. A. L. Bellardi of Turin for exploration of the high mountains. These were difficult trips enabling him to test himself, and they whetted his desire for more distant travel. Lemonnier then came through with funds that enabled him to visit Cyprus, Syria, and Lebanon in 1787-1789. He had just begun the publication of plants from that expedition in 1791, *Icones plantarum Syriae rariorum*, illustrated by P.-J. Redouté, when he was interrupted by the opportunity to accompany Entrecasteaux.[142] On this occasion, his name was advanced by Thouin and Jussieu, and the nearly-bankrupt state would be the patron.

Labillardière at once recognized the great opportunity for a naturalist and believed himself to be well-prepared for all the hardships inevitable on such long voyages. What is more, he found it *glorious* to be among those obligated to make

142. Auguste de Saint-Hilaire. 1834. "Discours prononcé le 10 janvier 1834 sur la tombe de M. de La Billardière." *Annales des sciences naturelles. Botanique.* Ser. 2, 1: 39-40; Auguste Chevalier. 1953b. "Un grand voyageur naturaliste normand, J.-J. La Billardière (1755-1834)." *Revue internationale de botanique appliquée et l'agriculture tropicale.* 32: 98-102; Stafleu. 1967. *Adanson, Labillardière, De Candolle*, pp. 16-21.

every possible search to bring back to their homeland those men who had done so much for their country.

As the preparation of the two ships was well advanced toward the end of August, 1791, General d'Entrecasteaux, as he would be called according to the custom of that day, notified him to come to Brest, which he reached on 10 September. Along with the other naturalists, Labillardière prowled through the town in search of materials for the preservation of their collections. After some difficulty, he put 22 reams of stout paper in a very large format on board. He noticed that the two flutes, each displacing about 500 tons, did carry weapons, but only the number and variety deemed sufficient to provide protection against any venture attempted by aborigines. The ships were ready to sail on the 25th, only awaiting favorable winds to leave port.[143]

143. Jacques-Julien Houtou de Labillardière. 1800. *Relation du voyage à la recherche de la Pérouse*, 2 vols. Paris: H. J. Jansen. 1: viii-3.

VIII
The Voyage of Entrecasteaux: Part 2 From Brest to Tasmania

The expedition sailed from Brest on the morning of 29 September 1791. On the next day, Entrecasteaux announced that he had been promoted to the rank of rear admiral, and both Auribeau and Huon to a captaincy. The harbor of Santa Cruz de Tenerife was to be the first port of call, and there they dropped anchor on 13 October 1791. While the crews were occupied taking on wine to replace that which had been consumed, and storing an abundance of fresh meat, vegetables, and grapes sent by the French consul, Fonspertuis, the naturalists developed a plan to climb the peak of Tenerife (Pic de Teyde) and to visit other high mountains on the island.

The French consul arranged for a party of mules, accompanied by conspicuously unenthusiastic guides; while Lt. Rossel, responsible for the headquarters table on the Recherche, ordered the cook to prepare an excellent salmon pâté for the trip into primitive country. It would become a unique contrast with the worm-eaten hardtack and cheese with which they would be entertained on most later field trips. While most of the plants encountered during the first day out (October 15th) had been collected and published earlier, Labillardière was pleased to find a species he had recently published in the second instalment (Decas secunda, July 1791) of his *Icones plantarum Syriae rariorum* with the name *Periploca angustifolia* Labill.[144]

Even though this field trip was not productive of new species, it provided Labillardière an early opportunity to gauge both the stamina and the commitment of his fellow naturalists. On the second day, as they climbed higher and encountered much volcanic débris, their progress was difficult and slower. When darkness came on, they found a resting place amidst large pieces of lava rock, their only shelter from the strong wind. Nature, not consulting the travelers' needs, had provided little wood; and the tiny fire they built could not prevent them from spending a bad night.

In the morning, Piron and Deschamps declined to go any higher. The rest of the small party from the *Recherche* went on to the summit, an elevation of 3,700m. For the return trip, most of the party favored the shortest route possible. Only

144. F. Markgraf, in recognizing this species as *Periploca laevigata* Aiton ssp. *angustifolia* (Labill.) Markgraf. 1971. *Bot. Jour. Linn.* Soc. 64: 375, incorrectly cited the type subspecies as from the Canary Islands.

Labillardière and the gardener, Delahaye, preferred to continue botanizing and, with some difficulty, persuaded one guide to remain with them. Despite not finding novelties, they examined highly interesting plants during the outing such as the fantastic *Euphorbia canariensis* L. and the broom *Cytisus fragrans* Lam., a shrub growing immediately below the volcano's cone. They subsequently learned that Claude Riche and Jean Blavier, naturalists from the Espérance, had undertaken to climb the peak a day later but were still distant from the summit when they found themselves unable to tolerate the rarified air. The entire episode provides us additional insight into explanations for later antagonisms.

The field trip had also become an occasion to contemplate the local religious climate, Spanish Catholicism, which our naturalists of the French Enlightenment instantly deplored. Upon reaching the town of Laguna, seat of a diocese, they were struck by its numerous convents and were informed that at least half of the town's inhabitants were monks. Judging from the children's evident preoccupation with nothing other than religion, Labillardière commented that the residents of the island had been immersed since birth in antiquated religious ideas; and that one could only pity such unfortunates upon whom fanaticism and monkish intolerance exercised such a powerful and dangerous dominion.

On the second day of the outing, the party had anticipated getting underway at an early hour, only to discover that it was a religious holiday. The guides refused to proceed before attending mass, and several of them felt compelled to attend three of them. Labillardière's impatience turned to anxiety when he learned that it would be highly unusual to be en route at all on such an important day. But the gods smiled, and they were underway again about 9 o'clock in the morning.

Labillardière further observed that "the multiplicity of religious practices introduced among the local inhabitants did not prevent some of their women from wandering, rosary in hand, in the presence of our sailors every time they went on shore, several of whom would have a long time to repent having allowed themselves to be seduced by so much charm. The discipline aboard British ships, several of which were anchored nearby, protected them from such calamities. Those navigators, well aware of the danger to sailors on shore, simply confined them to their ships. A captain will permit sailors to go ashore only on ship's business. Our commander has vowed to adhere to that rule during every future halt on the African coast."[145]

After ten days of refreshment in port, the expedition sailed again on 23 October 1791, setting a course for Cape Town between Africa and the Cape

145. Labillardière. 1800. *Relation*. 1: 9-30.

Verde Islands. The ensuing days were botanically uneventful; but to the extent that botanists were also naturalists, they took an interest in observing fish and seabirds. After a few days, tuna and bonita in great numbers began accompanying the ships, and crewmen on the *Espérance* became especially skilled in catching a good supply of fresh food. All the more important as they had come sadly to realize that the vegetables and fruits purchased in Tenerife were not keeping well because of the heat and humidity in the doldrums. By late November, after they had left the doldrums, contrary winds impeded progress, water currents pushed them westward, and they began to fear that dwindling water supplies might force them to seek an anchorage in Brazil.[146]

Entrecasteaux, meanwhile, had been notably strict about shipboard sanitation. Not only had his instructions from Fleurieu emphasized the critical need for cleanliness on long voyages; but, having taken great precautions to engage a crew free of disease, as La Pérouse had done before him, Entrecasteaux had seen his efforts partially compromised by self-indulgent crewmen in Tenerife. Regular fumigations of the ships were ordered to eliminate fetid odors. Sailors exposed to rain were required to change clothes after every quarter of duty (six hours). His ruminations about sanitation exposed a personal concern for the health of his crews.[147]

The maintenance of health on board was critically dependent upon an adequate supply of safe water, and, by late December, the daily ration per man had been reduced to one bottle of water. Labillardière made the following observations: "We were quite aware that our supply of water would not retain its purity through the stifling heat we were enduring. But it would have been difficult to know in advance, as we had on board very easy methods of purifying water before the daily distribution to each person, that these methods would not be fully employed.

"Water conserved on board during long passages undergoes the same decomposition as stagnant waters do, and such decomposition is especially accelerated by a hot climate. The water then gives off such a large amount of inflammable air that you run the risk of asphixiation by going down into the hold where the water is stored. Such an accident is quite unusual, as the opening leading into the hold allows some of the noxious fumes to escape. But it is also true that they frequently give birth to nervous disorders, whose malignancy is proportional to the degree of heat that decomposes the water.

146. Ibid. 1: 34-51.
147. Rossel. 1808. *Voyage d'Entrecasteaux*. 1: 14-18.

"As this gas, whose specific gravity Joseph Priestley was the first to make known, is much lighter than the air in the atmosphere, and which moreover has little adhesion to water, it is easy to separate it from water and return this beverage to its original purity. To accomplish that, it suffices to agitate the water for a quarter of an hour.

"We had a device on board that accomplished this goal perfectly: a large double-hectoliter tub. When it was filled three-quarters full with water, the water was stirred toward the middle by four large iron blades, put in motion with a crank and balance wheel. A vigorous agitation of the water relieved it of the inflammable gas, simultaneously reintroducing the pure air of which it had been partly deprived, and eliminating all previous infection. In a short time, such water did not differ from the best water."

Given the availability of such a device, why did their drinking water feature the same stench that wafted from the hold? Labillardière explained in a manner that illustrated the source of tensions on board: "The officer on duty, responsible for surveillance of the operation, usually abandoned it to the care of a sailor, who, soon bored by turning the crank, nearly always pronounced the water sufficiently agitated before it was, in fact, potable. It would have been more suitable to confide the surveillance to the chief-surgeon as several among us pointed out. For *he* was not indifferent to the general health on board or to whether this operation was well or poorly executed. Nevertheless, the officer on duty continued to be put in charge."[148] The criticism had to be aimed at Captain d'Auribeau.

Labillardière added, in all justice, that there was not one sick person on board despite the water rationing, and he acknowledged that an attempt was made to compensate through increased use of various antiscorbutics. He noted in particular, as salubrious and very agreeable, a drink made with eau-de-vie, vinegar, sugar, and water that was distributed daily to the crews toward the end of the passage to Cape Town. Fumigations were rigorously carried out twice a day, highly commendable precautions in Labillardière's opinion.[149]

Consequently, they entered Table Bay in reasonably good shape on 17 January 1792. After an inspection for contagious diseases by two Dutch health officers, they were free to go ashore at Cape Town on the 19th. General d'Entrecasteaux was confronted immediately by a dilemma. The French chargé d'affaires handed him a dispatch from Governor Saint-Félix, the French commander of the Indian Ocean station on Mauritius. The dispatch was, in fact, a covering letter for deposi-

148. Labillardière. 1800. *Relation*. 1: 55-57.
149. Ibid. 1: 64.

tions made by the captains of two French merchant ships after they had reached Mauritius from Batavia (Djakarta). The first deposition made to Governor Saint-Félix had been given on 6 October 1791 by Captain Préaudet commanding the *Jason*. The second deposition, dated 31 October 1791, had been made by Captain Pierre Magon de l'Epinay commanding the *Marie-Hélène*. Both depositions were reports of conversations the captains had allegedly had in Batavia with Captain John Hunter, commander of the British frigate *Syrius*, that had been shipwrecked on Norfolk Island east of Australia.

Hunter and his staff had been rescued by a British corvette and put ashore at Botany Bay. He had then contracted with the commander of a Dutch ship, the *Waaksamheyd*, to take them back to Europe, a long trip going via Batavia. The Préaudet deposition conveyed the critical information that Hunter had been at Botany Bay on an earlier occasion in 1788, when La Pérouse had been refitting there. He had learned directly from La Pérouse what his itinerary would be and, in particular, knew that La Pérouse meant to pass through the Saint George Strait, which would have taken him into what we now call the Bismarck Archipelago and into the proximity of the Admiralty Islands. That testimony was confirmed by Pierre Magon's deposition.

The *Waaksamheyd*, on its passage from Botany Bay to Batavia, had been beset by contrary winds and currents, carrying the ship farther east from its projected route. As a consequence, Hunter had inadvertently passed through the Saint George Strait, according to the depositions; and, near the Admiralty Islands, natives in pirogues had been seen wearing materials of European origin, seemingly uniforms and sword-belts of French naval soldiers. The British observers presumed these must be the remains of La Pérouse's crews.

Governor Saint-Félix was sufficiently impressed by the evident integrity of the two French captains that he felt obliged to transmit their depositions to Entrecasteaux. Having been informed officially of Entrecasteaux's assigned itinerary, he knew that Entrecasteaux would be stopping on Mauritius only on his return trip. Therefore, he dispatched the frigate *Attalante* under Captain Bolle to Cape Town so that this critical information could be conveyed to Entrecasteaux when he arrived. Bolle had then returned immediately to Mauritius and was unavailable for questioning when Entrecasteaux was given the information.[150]

That information should have been a godsend. Instead, it unleashed doubts. Why, to begin with, would La Pérouse have told Captain Hunter of his intention to pass through the Saint George Strait when such a direction was incompatible with the route he had specified in detail in his last report to the ministry of the

150. Rossel. 1808. *Voyage d'Entrecasteaux*. 1: 19-25; Labillardière. 1800. *Relation*. 1: 66-73.

marine from Botany Bay? Why were there evident discrepancies, if minor, in the descriptions of European clothing as reported in the two depositions? Indeed, why had not Captain Hunter, given his suspicions, persuaded the Dutch captain to reconnoiter closer to shore for possible refugees? A sacred duty as Entrecasteaux put it! To make matters more confusing, Entrecasteaux soon discovered that Captain Hunter had actually been in Cape Town on his way to England when the *Attalante* had brought the critical news from Mauritius. The word had quickly spread through the waterfront people, Hunter had heard it, and he told several members of the regency of the Dutch East India Co. that he knew nothing about the facts being attributed to him.

That had to raise the appalling possibility that the two French captains had simply reported, as fact, waterfront gossip in Batavia that had been embellished from mouth to mouth. Why, then, had not Captain Hunter seized the opportunity to clarify the record as he was still within the roads of Cape Town when two French ships, one flying the flag of an admiral, sailed into the harbor? There should have been no doubt that it was Entrecasteaux. Instead, Hunter set sail for Europe two hours later and was long out of reach when the French discovered the urgent need to consult him. Did that mean that Hunter had seen nothing to report? Or that he had belatedly recognized the possibility of censure for having failed to press the search for refugees and sought escape through denial?[151]

During the several weeks needed for victualing and refitting at Cape Town, Entrecasteaux had time to ponder whether to adhere to the prescribed itinerary or to exercize his latitude by setting the most direct course for the Admiralty Islands. He needed that time, moreover, to cope with a written request to resign from the expedition that bore the names of Labillardière, Riche, Delahaye, Blavier, Piron, and Chailly-Ely. When Entrecasteaux sought to understand the reasons for dissatisfaction, Claude Riche told him that they had come to fear their personal expenses would never be reimbursed. Beyond which they believed that the private collections being made by the medical officers on board either ought to be forbidden or should be turned over to the naturalists to become part of the official collections.

While Entrecasteaux gave assurances that reimbursements would be made, and reminded Riche that the king's instructions required that *all* collections were to be submitted for identification and description by the four official naturalists as part of the government's collections, he remained far from convinced that he had been given the truth. Even the solidarity of the signatories was suspect.

151. Rossel. 1808. *Voyage d'Entrecasteaux.* 1: 26-31.

He had, for instance, obtained horses and guides to facilitate fieldwork around Cape Town and knew that Labillardière, Riche, and Delahaye were pleased to be making collections of plants and seeds. They hardly seemed disaffected.[152] Stranger still was the absence of the abbé Claude Bertrand's name from the list. An able astronomer but cursed with weak eyes, Bertrand had been unable to do the work of a shipboard astronomer. He had also revealed himself, from the time the expedition had left Tenerife, to be a selfish troublemaker and had been suspected of scheming to engineer a mass resignation. His defective eyesight enabled Entrecasteaux to give him a medical discharge. He applied the same solution to Jean Blavier, the mineralogist, who had suffered cardiac palpitations when climbing on Tenerife; as well as to the painter Chailly-Ely whose abundant ailments hinted at a pathological fear of the sea.

Entrecasteaux's diagnosis of the uprising, and his management of it, indicated a shrewd insight into personnel and a judicious temper. An able administrator early learns to discount the reasons cited for discontent or a resignation and applies himself to discern what is not being said. In this instance, by making a judgment based upon prior and current experience, he rid himself of troublesome naturalists in a politic way and retained those likely to accomplish their expected work. Neither Deschamps nor Ventenat had signed the request to resign. Neither had they shown an eagerness to explore beyond Cape Town as had Labillardière, Riche, and Delahaye.

Entrecasteaux understood, as had La Pérouse before him, that, whatever might be individual gripes, all savants had a high opinion of themselves and suffered from the corrosive suspicion that they were not appropriately respected by the naval officers.[153] That was an ailment beyond the reach of any drug in the pharmacopeia. The conflict would only be resolved later in the 19th century when the French navy would train its own officers for scientific work on expeditions.

A final curious aspect of that incident reported by Entrecasteaux is that Labillardière made no mention of it except to express regret over the resignation of three colleagues. His complaint of the moment concerned the fate of 400 Blacks who had recently been landed in Cape Town from a slave ship from Mozambique and were being warehoused in three impossibly crowded rooms pending transshipment to a market in the Americas. Most of them were wretchedly scorbutic, after only a brief stretch at sea, having been deprived of their traditional food. The slavers, coming from Mozambique, were likely either Portuguese or Arabic

152. Ibid. 1: 32-33.
153. Horner. 1995. *Looking for La Pérouse*, pp. 55-57; Labillardière. 1800. *Relation*. 1: 97.

and had no qualms about admitting that they had often been able to procure two or three Blacks in exchange for one good dog. His descriptive paragraph bristled with disapproval.

After Labillardière's long confinement at sea, the opportunity to botanize in new country was a liberation; and, despite the fact that it was hardly *terra incognita* to European botanists by the end of the 18th century, he spent his days on Lion Head, Devil's Peak, Table Mountain, and examining the hills around Stellenbosch when pushing toward the Franche-Hoek Mountains. His little party occasionally encountered runaway slaves who had sought refuge in the mountains behind Cape Town to escape barbarous treatment. Under cover of night, the runaways approached habitations in a search for food, although at peril to their lives.

He made abundant collections in the vicinity of Cape Town. But when he took time to determine and preserve his specimens, and whenever he enjoyed the companionship of Francis Masson, a gardener from Kew who collected overseas for Joseph Banks, Labillardière was forced to recognize that he was gathering species known to Europe. That is what drove him toward the Franche-Hoek Mountains by the end of January, 1792, with the company of Delahaye.

Once they got inland, the place-names indicated an area settled by French Protestant refugees in the 17th century. Such original settlers, required to speak Dutch, had preserved their maternal tongue only in their paternal names, such as one of their hosts, Jacob de Villiers. Labillardière also observed that the inland farmers, although giving the most arduous tasks to their Black slaves, generally treated them gently. He found it remarkable that the Dutch, unlike the Spanish who had always endeavored to make converts, left their slaves in total ignorance of the Dutch religion.

As the expedition was scheduled to set sail in mid-February, Labillardière and Delahaye were obliged to cut short their visit inland, and they walked for about 24 hours to reach the *Recherche*. Given the overcrowding on the resupplied ships, Entrecasteaux induced Labillardière to turn over his South African collections to the French chargé d'affaires at the Cape, who promised to have them shipped back to France at the earliest opportunity. Labillardière also notified Thouin by letter that Delahaye would be sending a substantial quantity of seeds he had collected packed in soil mixed with sand.[154]

Not until Labillardière reached Mauritius on his way home in 1795 did he learn that the chargé d'affaires had not fulfilled his pledge in 1792, despite many opportunities to comply. Aubert du Petit-Thouars, by then on Mauritius, told

154. Letouzey. 1989. *Le Jardin des Plantes*, pp. 239-240.

Labillardière he had seen the collection relegated to the French agent's storehouse on the Cape long after Labillardière's departure. Clearly they had long since been forgotten, and he would never see them again.[155]

Beyond the loss of valuable collections, the incident is a reminder of the easily overlooked peril seafarers routinely experienced because of their vital dependence upon the integrity of their providers in ports of call. Purveyors in general were not activated by a spirit of charity and were geographically placed to enjoy an enviable monopoly. What is more, third-rate merchandise, concealed with first-rate prices, may not reveal its truth until weeks later at sea when it is not only too late to seek redress but a threat to daily life. The ultimate calamity for a captain was to have on board an intelligent and witty savant (Labillardière) who could comment mercilessly on the capacity of naval officers to be victimized by slick entrepreneurs.

In private, Labillardière quite recognized that the mercenary spirit of the Dutch company, which had a chartered monopoly to govern the Cape of Good Hope, left its customers little option but to submit to exorbitant prices. He had also heard naval officers debate the wisdom of switching to St. Helena where they could resupply at lower prices. But St. Helena did not possess the bountiful hinterland that had always made Cape Town such a logical port of call. So, Labillardière wondered, how had it happened that the governments of Europe had been so ignorant as to have left one of the most important places on the sea route to the Indies for so long in the hands of a company of merchants, leaving the various national interests subordinate to one company's concern for profits?

In truth, the situation was all the worse for Entrecasteaux in that, at Cape Town, he had had to replace commodities originally taken on board that had proved to be not only of poor quality but deceptively advertised in Brest. He had paid, for instance, double the usual price for wine in order to obtain a quality that would keep for a long time. A portion of it had spoiled before he reached Cape Town. He now had a choice between a dubious local wine at a daunting price or some Bordeaux on a trading ship in the harbor. Given their projected route, there would be no alternative elsewhere to procure potable wine. For reasons incomprehensible to Labillardière, Entrecasteaux failed to choose between his two options, and the ships would sail from Cape Town on 16 February 1792 with their own bad wine that got increasingly sour. In Labillardière's medical opinion, good wine, with its powerful preservative agent, could contribute to maintaining health during a long voyage where one would be exposed to a variety of privations. The only substitute would be increased dependence on eau-de-vie.[156]

155. Labillardière. 1800. *Relation.* 1: 77-98.
156. Ibid. 1: 99-100.

As for his other pressing options, the general had decided he must rely on the depositions of the two French captains, perhaps because he feared criticism if he failed to seize the opportunity to find La Pérouse more expeditiously than by adhering to the prescribed route. He cannot have been unaware of the navigational difficulties he would face if La Pérouse should not be found on the Admiralty Islands. The shortest route required passing along the north shore of New Guinea, which he would have to reach before the annual reversal of the monsoon winds.

After 21 days of violent storms and heavy seas in the Mozambique Channel, it became obvious that his "bad sailors," the makeshift frigates, would not get beyond Timor before the onset of the monsoon from the east. Captain Huon agreed with his decision to turn southeast to resume the prescribed itinerary. They sighted Amsterdam Island in the South Indian Ocean on 22 March 1792, roughly the midway point between the Cape of Good Hope and Tasmania; but, conscious of the loss of time and their slow progress, they made no attempt to land.[157]

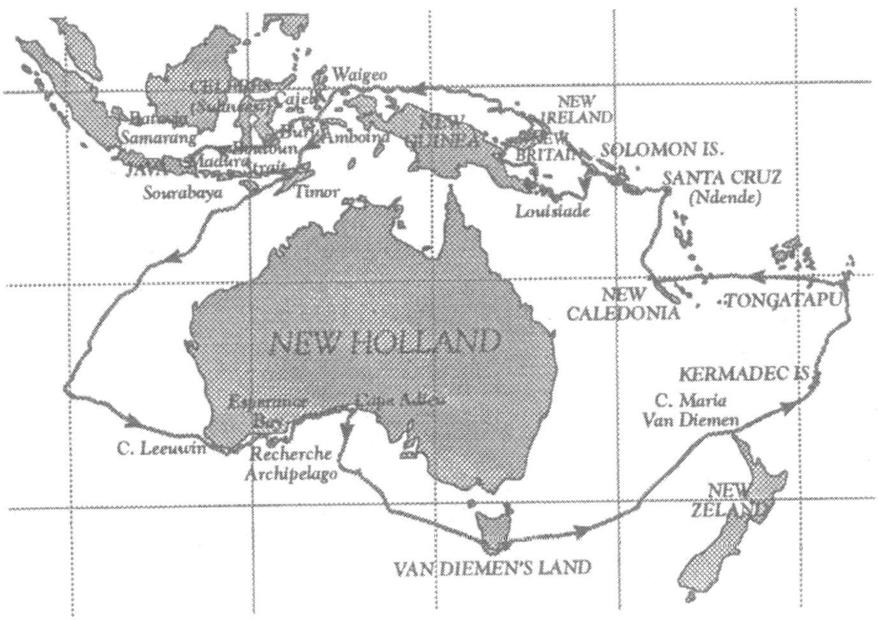

Entrecasteaux's track in 1792-1793

157. Ibid. 1: 105-112; Rossel. 1808. *Voyage d'Entrecasteaux*. 1: 38-45.

When loading supplies at Brest, the naval officers believed they had been sold fresh, high-quality hardtack. By the first week of April, they realized that at least part of their supply had been on a previous campaign. After five months of navigation, their hardtack was inhabited by a prodigious number of larva that were producing small beetles of the species *Dermestes panicens*. In the evenings, the insects would be attracted to their lights in such numbers as to extinguish them. They also invaded the rest of the food, drawing from Labillardière the remark that a good deal of time had to pass before the repugnance the insects had awakened began to lessen. The high winds and frightful waves encountered during the crossing of the South Australian Basin were trouble enough without the bugs.[158]

Consequently, Entrecasteaux was greatly in need of a secure anchorage that would allow time for repairs from wind damage. His initial target was Adventure Bay, charted on the southeast coast of Tasmania, that would offer protection from the prevailing west wind. For reasons unclear, because of confusing testimony, the expedition entered an attractive bay, 20-21 April 1792, after rounding the south cape of Tasmania on the assumption it was Adventure Bay. But their inability to locate Penguin Island, charted within Adventure Bay, forced them to recognize that they had entered an unmapped bay, but one that could well serve their urgent needs.

Because that Tasmanian location, reached inadvertently, would become one of the major sites for botanical exploration during the entirety of the expedition, we must remark about place names the explorers conferred upon the region. The mapping was done by the engineer-hydrographers, especially Charles-François Beautemps-Beaupré, in 1792 and during a second visit to the bay in 1793. The maps reveal, first of all, that Adventure Bay was not an indentation on the mainland coast as they had anticipated, but rather an indentation on the east side of a long, offshore island that they would name Bruny Island. The long, narrow strait between that island and the mainland they would name Entrecasteaux Channel, both names surviving today.

As for the bay they had actually entered, they named it the Recherche Bay. Within it they would find two exceptionally convenient anchorages close to shore, designated as North Port and South Port.[159] The initial reconnoitering of the bay was carried out by two officers in a small boat. North Port was selected as their initial harbor as a good stream of water flowed into the sea at that point, 23 April 1792. The harbor was found to contain great quantities of fish, easily obtained by

158. Labillardière. 1800. *Relation.* 1: 113-116.
159. Rossel. 1808. *Voyage d'Entrecasteaux.* 3 [1807]: Plate no. 3.

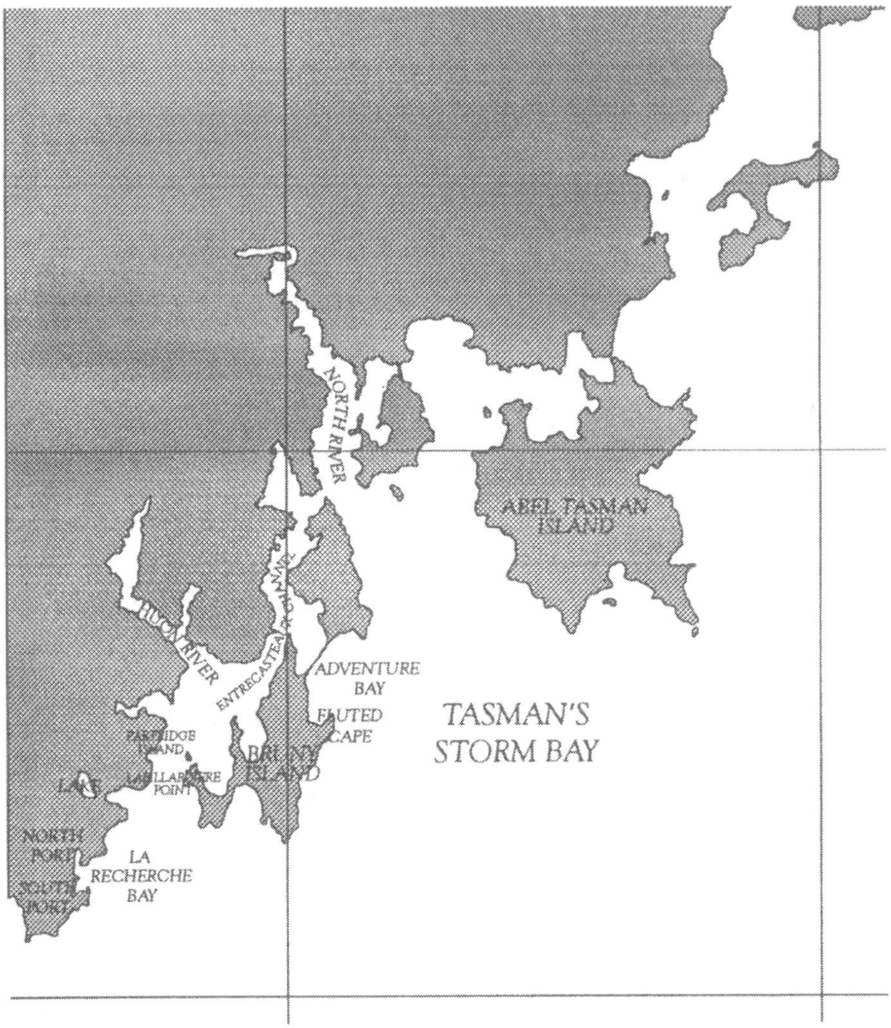

lowering nets or on a line, an immediate source of fresh food, the most numerous being a species of *Gadus*, codfish.[160]

Crude shelters constructed from tree-bark were scattered along shore, and smoke from a fire at a considerable distance left no doubt that aborigines were living nearby. Labillardière and Delahaye, escorted by two crewmen, went ashore that very afternoon to explore toward the northeast. "We were struck with admiration," he recorded, "by the sight of those ancient forests that the axe had not yet

160. Ibid. 1: 47-60.

degraded. The eye was astounded by the immense height of those trees. Some in the Myrtle Family were more than 150' tall, their tufted tops crowned with an evergreen foliage. Several, falling from decrepitude, leaned against their neighbors and were brought to earth only as they destroyed themselves. The most vigorous vegetation provided a striking contrast with that state of wasting away, and you could see in all its magnificence the imposing tableau of Nature, which, left to itself, only destroys to recompose."[161]

Collections of various *Eucalyptus* were made on that outing, including *Eucalyptus resinfera* Sm. in White; as well as a new *Epacris*. We cannot tell which new heath it was as Labillardière would publish five new *Epacris* from Tasmania without collection dates in *Novae Hollandiae plantarum specimen*, a title that serves to remind that in his time Tasmania was still regarded to be the southeastern tip of Australia.

On the following day, 24 April 1792, he worked until mid-morning to describe and prepare his collections on the 23rd, then went ashore again to explore to the east of their anchorage. Several attempts to push inland were frustrated after hardly a kilometer by the density of the forest, forcing him back to the beach. Not only the shrubs prohibited entry, but passages were frequently blocked by fallen tall trees.

The various species of *Eucalyptus* struck him as outstandingly handsome trees. Many of the large trees near the sea had been hollowed at their base by fire, always on the lee, by natives to provide protection from the wind while eating. The residue of crustaceans and shellfish revealed their diet, and ashes indicated that fires had been lighted for cooking, the food prepared on coals. When following the shoreline to the southeast, they attempted to cross a boggy area, utilizing spots strengthened by plant roots; but a new species of what is now called sword-sedge cut both hands and faces and forced a retreat to the harbor.[162]

On the 25th, after giving Piron a few plants to draw, Labillardière and Delahaye set out again to the southeast along the shoreline. Reaching an opening into the forest, they soon came upon a windbreak constructed by natives using bark from *Eucalyptus resinifera*, intertwined among stakes driven perpendicularly into the ground: arc-shaped, one-third the circumferance of a circle 3m wide, 1m high, the convex side facing the sea. This provided protection for the cooking-coals, and the debris from shellfish was evident. Later, they found a partially-built hut as they were returning through the forest to the ship. Grasses had been used as

161. Labillardière. 1800. *Relation*. 1: 128.
162. *Lepidosperma elatior* Labill. 1804. *Nov. Holl. Pl.* 1: 15. Tab. 11 (1805).

ties to consolidate the framework and to fasten the eucalyptus bark that made the hut impervious to rain. But, no one as yet had encountered a single inhabitant.

Labillardière devoted 26 April 1792 to preparing and describing specimens; but at daybreak, Delahaye and he went ashore meaning to have a long day. Pushing southeast, they encountered a pathway obviously used by natives and immediately followed it to the south. He was delighted to find a species of *Erigeron* with a woody stalk covered by very small fleshy leaves.[163]

The botanizing was interrupted when a member of their escort fired a shot meant to bring down a bird. A frightened young native boy, entirely nude, was spotted running into the thickets. Everyone ran after him in the hope of an interview, but the search was fruitless. The hope of meeting additional aborigines led them to push on into the forest with the intention of spending a night out. After an hour beating a difficult track to the southeast, they reached a broad plain that extended to the sea, and Labillardière was struck by what he thought was a fine mimosa growing there: leaves long, ovate, entire, their longitudinal nerves prominent; legumes semicircular; height 8-10m.[164]

Because the party of seven had carried provisions for only a day's outing, rations were reduced that night in order to provide for the next day. Boughs were cut with a battle axe to build a shelter, and a local fern was used to soften the ground.[165] The cold soon obliged them to abandon the hut and to sleep near their fire. With daylight, they went hunting in search of breakfast, and two members of the party soon brought in a crow and an oystercatcher. The birds were grilled on the spot and eaten as if they had been delectable food. Only then was it perceived that the provisions for the day had been confided to unreliable hands. Of the six cakes of hardtack, only four remained. This betrayal meant an immediate return to the ships following the shoreline.

During the return, Labillardière found a new species of parsley whose stalks lay on the ground, hence the name he gave it.[166] Analogy with known species in the same genus led him to suppose that it would be a good food, so an ample supply was collected for the ship where it was received joyfully by navigators in need of plants to overpower the effects of long dependence upon salted meats. Among other plants collected on the return was a sundew distinguishable by its singular leaf-form: two long blades at the end of each petiole that emerged from the plant's root.[167] He also collected several species within a new genus that had

163. *Erigeron pappocroma* Labill. Ibid. 1806. 2: 47. Tab. 193 (1806).
164. *Mimosa sophorae* Labill. Ibid. 1806. 2: 87-88. Tab. 237 (1807). SY=*Acacia longifolia* Willd.
165. *Polypodium scandens* Labill. Ibid. 1806. 2: 91-92. Tab. 240 (1807). SY=*Polypodium diversifolium* Willd.
166. *Apium prostratum* Labill. Ibid. 1804. 1: 76-77. Tab 103 (1805).
167. *Drosera binata* Labill. Ibid. 1804. 1: 78. Tab. 105 (1805); Labillardière. 1800. *Relation*. 1: 129-143.

great affinity to *Polygala* L., presumably *Comesperma*.168 For the next two days, he spent most of his time describing and preparing his abundant harvest.

The meals being served on board, thanks to the daily netting of an abundant supply of fish, were a striking contrast to what the naturalists had been obliged to tolerate during excursions on shore. "It is well to remark," Labillardière observed, "that the naturalists, having asked for fresh foods to carry on their outings, for which they had a right, obtained absolutely none. Hardtack, cheese, eau-de-vie, and on occasion some salt pork, was our usual provisioning and, despite our rights to better fare, would continue to be our lot throughout the campaign. I would abstain from reporting this fact if I did not believe it could be of some use to naturalists who will undertake voyages in the future."[169]

1 May 1792: Landing on the western side of the harbor, Labillarière's party moved northward, occasionally turning aside into the forest. He was pleasantly surprised to find a new genus of mushroom emerging from the mosses. The arrangement of its rays led him to name it *Aseroe*: "Its roots are small filaments attached to a fungous tubercle, on which rests a volva, globular, whitish, gelatinous, marked with seven stria inside and outside. From the middle of this volva, a reddish peduncle (stipe) rises, nearly cylindrical, grooved its entire length and open at its upper extremity which is flared, with a fine red color, and divided into seven bifurcated rays, yellowish at their extremity. This mushroom is smooth in all parts. The new genus must be placed adjacent to the Linnaean genus *Phallus*."[170]

The mountains to the west could be penetrated through a beautiful valley where frequent springs were uniting into small streams that flowed down into the harbor. The shrubbery was dense and the area dangerously boggy. A new species within the Cyperaceae was an even more fearful barrier. Reaching a height of near 4m, its leaves were as sharp as a piece of glass. Its oval, reddish seeds, containing a small, oily kernel, were greatly favored by parrots despite the hard shell.[171] The most common shrub in the bogs was an *Embothrium*, remarkable for the toughness of its foliage: each leaf oval in appearance, but 1dm long and 3cm wide.[172]

On May 3rd, Labillardière led a party farther northward than they had previously visited, reaching a large lake. Once again there were signs of natives, attracted to an abundance of shellfish, but no natives were visible. Plant collecting was

168. *Comesperma* Labill. 1806. *Nov. Holl. Pl.* 2: 21.

169. Labillardière. 1800. *Relation*. 1: 143-144.

170. Ibid. 1; 145. For the type species, see *Aseroe rubra* Labill. Ibid. Tab. 12.

171. *Lepidosperma elatior* Labill. 1804. *Nov. Holl. Pl.* 1: 15. Tab. 11 (1805).

172. *Embothrium truncatum* Labill. 1804. Ibid. 1: 32. Tab. 44 (1805). SY=*Telopea truncata* (Labill.) R. Br.; Labillardière. 1800. *Relation*. 1: 145-146.

highly successful, and he mentioned in particular a mimosa with simple leaves, actually an acacia;[173] and an especially striking member of the Lily Family notable for brilliant red flowers.[174]

The burgeoning backlog of specimens, both plants and bird skins, forced Labillardière to devote his next two days to their description and preservation. The very limited space he occupied on board forced him to use an oven to promote the drying of his paper-driers after use. And, once his room was fully occupied, he had no other place than the main saloon to lay out additional plant specimens until they would be entirely dry. Captain d'Auribeau, commander of the *Recherche*, determined that natural history items must not be allowed to intrude upon that locale, even temporarily, and had two of Labilladière's plant presses put outside. Labillardière appealed the matter to the commander-in-chief; and Entrecasteaux, who must have been impressed by Labillardière's evident dedication to his assignment, gave orders that the displaced presses should be returned to the saloon.[175]

Among the various eucalyptus Labillardière had observed during his botanizing, one intrigued him in particular because of its potential utility. One of the tallest trees in nature, and with a close affinity to *Eucalyptus resinifera* Sm. in White, its fruit had been found on the ground; but the flowers were near the summit of the tree. He believed that the trunk could be suitable for naval construction; and, even though not as light or flexible as pine, could be used for masts. Could the trunks not be cored throughout their length to make them lighter, then strengthening them with iron rings at intervals?

On 6 May 1792, he cut down one of the trees, already leaning, to collect its flowers. Because the fruit had the appearance of a clothing button, he gave the new species the name *Eucalyptus globulus*. He surmised further that, as the bark, leaves, and fruit were aromatic, they might have economic value in a world so long dependent upon the Moluccas for its aromatics.[176]

Having covered the area in the vicinity of the anchorage, Labillardière left with a small party on the afternoon of 8 May 1792 with the intention of spending three or four days in the field to extend his range. He went northward to the vicinity of the large lake previously partially explored. After one night out, the painter, Piron, returned to the ship. Shortly thereafter Labillardière found a new evergreen tree whose kernel, like that of the cashew nut, was situated in a much larger fleshy receptacle. For that reason, he named the new genus *Exocarpos*. The

173. Probably *Mimosa suaveolens* Labill. 1806. *Nov. Holl. Pl.* 2: 87. Tab. 236 (1807). SY=*Acacia pycnantha* Bentham.

174. *Aletris punicea* Labill. 1804. Ibid. 1: 85. Tab. 111 (1805). SY=*Blandfordia marginata* Herb. in Lindl.

175. Labillardière. 1800. *Relation*. 1: 149.

176. *Eucalyptus globulus* Labill. Ibid. 1: 153. Tab. 13.

plant had bisexual flowers as well as distinct male and female flowers. The principal characters showed him great affinity to *Anacardium* L. and led him to place it in Terebinthaceae. (Robert Brown would transfer it to Santalaceae.) The type species was *Exocarpos cypressiformis* Labill.[177]

At nightfall, they settled down on a stream bank surrounded by delightful groves of a species he named *Thesium drupaceum* Labill., a species with narrow leaves although his illustration did not show leaves.[178] The cold obliged them to light a large fire. Just as they were falling asleep, they were startled by the cry of a fierce beast nearby, comparable to the cry of a leopard. Several days before Labillardière had found the upper jaw of a fair-sized, four-footed carnivore, possibly a Tasmanian devil. The fire, he presumed, was serving a double purpose.

10 May 1792: Up with daylight, they continued along the lakeshore, encountering two relatively new native huts, but without natives. The great find of the day was a beautiful plant distinct from anything yet described: an iris with only two stamens. Because of that singularity, he contrived the genetic name *Diplarrena* (or *Diplarrhena* as corrected by Robert Brown); and, given its affinity to the genus *Moraea* Mill. ex L., he designated the species as *Diplarrena moraea* Labill.[179]

The flowers were extraordinarily delicate, fading much more quickly than other genera in the Iridaceae, and he would have given up all hope of having the species drawn had he not found additional plants after the initial collection. The spathe was subtended by 2 valves (bracts). The corolla had 6 divisions, the 3 interior being much smaller than the 3 exterior. Although he did not recognize a calyx, it was, in fact, more conventional by 1792 to characterize the Iris Family as having 3 sepals and 3 petals.

"Having examined a large number of flowers," he explained, "I have consistently found only 2 stamens whose filaments bear white-colored anthers marked with 2 grooves. In the place where there should be a third stamen, there is always a rudimentary filament not more than 2mm long that bears no anther. The ovary is inferior, has three angles, and is borne on a long peduncle. The style, somewhat longer than the stamens, is cylindrical and terminates with a stigma shaped like a shepherd's crook. Each of the 3 locules in the capsule contains several globular seeds."

On 11 May 1792, they bore westward, crossing a wide plain, dotted with marshes so covered with plants that they concealed the dangers underneath. Given the evident perils and the fact that provisions were running low, it seemed wise

177. *Exocarpos cypressiformis* Labill. Ibid. 1: 155-156. Tab. 14.
178. *Thesium drupaceum* Labill. Ibid. 1: 156; Labillardière. 1804. *Nov. Holl. Pl.* 1: 68. Tab. 93. SY=*Leptomeria billardieri* R. Br.
179. *Diplarrena moraea* Labill. 1800. *Relation* 1: 157-159. Tab. 15.

to return to the ships before dark. The entire next day barely sufficed to describe and prepare the collections of the preceding days.

Aside from the new ***Apium***, the botanizing had not disclosed any notable supply of edible plants. They had on board a great variety of European seeds, and the temperature range they had experienced around Recherche Bay gave hope that the seeds could succeed there. But what would be the use? Not only were the elusive aborigines unavailable for horticultural instruction, but their campsites suggested a diet quite devoid of plants. Even though Entrecasteaux assumed that the natives were too primitive to understand the properties of garden plants, he agreed that Delahaye should plant various seeds that could provide valuable foods for anyone else who might visit their harbor. Consequently, Delahaye had been busy spading and planting a garden quite close to the east side of the harbor.[180]

Any discussions of native diet could not avoid reference to a discovery made by Dr. Riche. He had found human bones in the ashes of a fire that had been built by aborigines, recognizing the pelvic bones of a young girl. They were only partially covered by burned flesh. While the possibility of cannibalism had to be admitted, Labillardière preferred to presume that the aborigines were in the habit of burning their dead and would not do them the injury of classifying them as cannibals on the basis of one such find.[181]

Beginning in late April, Entrecasteaux had dispatched longboats carrying his engineer-hydrographers, Beautemps-Beaupré and Jouvency, to reconnoiter the coastlines around Storm Bay and to check water depths. It gradually became clear that the channel, from which the ships had originally entered Recherche Bay, had every appearance of being a strait extending to the northeast and offering good anchorages throughout. It was given the name Entrecasteaux Channel. In the process, they recognized that the land to the east, initially believed to be a cape, was actually an island, henceforth Bruny Island.[182]

By 14 May 1792, repairs to the frigates were virtually completed so that an attempt to sail through the channel was imminent. Labillardière and Delahaye went ashore for one last outing, initially to visit the latter's garden. Labillardière found it well-spaded and arranged in four sections, a plot 9m x 7m; but the clay soil made him doubt any great success. Moving beyond into an area new to them, they were assailed at midday by a downpour, and the day ended in a nightmare of steep slopes and fallen timber, but with a notable new species.[183] They found the two frigates prepared to sail.

180. Ibid. 1: 154; Rossel. 1808. *Voyage d'Entrecasteaux* 1: 63.

181. Labillardière. 1800. *Relation*. 1: 161-162.

182. Rossel. 1808. *Voyage d'Entrecasteaux*. 1: 78.

183. Labillardière. 1800. *Relation*. 1: 163-165; *Restio tetraphyllus* Labill. 1806. *Nov. Holl. Pl.* 2: 77-78. Tabs. 226-227.

At sunrise on 16 May 1792, the wind became favorable for leaving Recherche Bay, and the turned northward meaning to enter the Entrecasteaux Channel. Almost at once their progress was stymied by a contrary tide and light north wind. In all directions, one could see spacious coves where navigators could expect to find shelter. Labillardière took advantage of the deadlock on the 19th to take a longboat to shore on Bruny Island (to an area still known as Labillardière Peninsula) and collected many plants unknown to him.

Most species belonged to *Melaleuca* L., *Epacris* Cav., and what he thought to be *Aster* L., but which were within *Olearia* Moench, a genus that replaces *Aster* in that region of the world. In the course of the day, he met Claude Riche who had also come ashore from the *Espérance* to botanize. They encountered occasional huts and piles of seashells similar to those on the mainland indicating that the island was inhabited.

On the 20th, both botanists descended upon a small island adjacent to Labillardière Peninsula that earlier navigators had named Partridge Island. Instead of partridges, they found an abundance of quail. Labillardière's new parsley, *Apium prostratum*, was abundant, growing along the shore near the high-water level, which they gathered in quantity to take on board, not failing to contrast their care to provide fresh food with the established policy of limiting the naturalists to hardtack and cheese when on shore.

That same day, two officers from the *Recherche*, Auribeau and Crétin, were reconnoitering the coast of Bruny Island. Noticing smoke from several fires close to shore, they landed in the hope of meeting aborigines. Almost immediately they came upon four natives cooking on three fires. They fled at once, leaving behind lobsters and shellfish on the coals. But very soon, a tall and strongly-muscled man returned to retrieve a basket of flints, seemingly unafraid, a confidence perhaps deriving from his evident physical strength. He came quite close to Lt. Crétin. This encounter provided the first physical description anyone on the expedition had been able to make of the natives. They were dark-skinned, but not deeply so; they had woolly hair, and the men let their beards grow. Several had kangaroo skins over their shoulders, but were otherwise entirely nude.

They had left behind nearly thirty rush baskets, about 30cm tall, crudely made from a species similar to *Juncus acutus* L., mostly containing shellfish and crabs. Smaller baskets, about 10cm, held pieces of flint bound with a soft tree-bark. Labillardière believed that striking flints together to obtain fire was a method unknown elsewhere in the South Seas or eastern Australia. The two officers prevented the sailors from carrying off any of the aborigines' property, taking only one basket of each size and a kangaroo skin for the General, but leaving in exchange several knives, several handerchiefs, and some hardtack and cheese.

Labillardière spent the next day, the 21st, preparing specimens. But Delahaye went back to the site in the hope of visiting natives. Again they would not allow themselves to be approached. It appeared that they had accepted the hardtack, but the cheese was untouched. Aparently its odor had prevented them from tasting it.

Meanwhile, a longboat sent out from the *Espérance* toward the mainland had found a cove into which a delightful stream of good water flowed. A detail was ordered to draw fresh supplies on the 22 May 1792, and Labillardière took advantage of the opportunity to collect in that locale. Among several species of *Casuarina* Adans. he found on the streambank, he thought one was notable because of its club-shaped fruit.[184]

Favorable winds enabled them to weigh anchor on 25 May 1792 to proceed through the channel, but adverse currents halted them again on the 27th. Labillardière again went ashore on Bruny Island, its more northern part. Among the handsome plants he collected, he thought a fine composite would form a new genus; and he would later name it *Richea* to honor his colleague on the expedition. The glaucous color of the leaves dictated the specific name.[185] Unfortunately, the genus would not stand, but his intent would be confirmed.

The unfavorable winds had inadvertently facilitated a more extensive exploration of the channel than Entrecasteaux had anticipated. The lateness of the season made it imperative to leave Tasmania for warmer climes. On the morning of 28 May 1792, when at least the tide was favorable, the expedition weighed anchors. Reaching the extremity of the island, and perceiving a large opening to the east-southeast, they sailed into the open sea and steered for New Caledonia.[186]

184. *Casuarina quadrivalvis* Labill. *Nov. Holl. Pl.* 1806. 2: 67-68. Tab. 218. SY=*Casuarina stricta* Dryand. in Ait.; Labillardière. 1800. *Relation.* 1: 173-181.

185. *Richea glauca* Labill. 1800. *Relation.* 1: 187. Tab. 16. SY=*Craspedia richea* Cass.

186. Ibid. 1: 189-190; Rossel. 1808. *Voyage d'Entrecasteaux.* 1: 92-102.

IX
The Voyage of Entrecasteaux: Part 3
New Caledonia to Ambon

The open sea proved to be extremely rough, and Labillardière found, after a few weeks on land, that he had lost his immunity to seasickness. Only on 3 June 1792 did the sea become sufficiently calm to allow him to get to work on specimens collected during his final outings. The naturalists were eager to reach the southern portion of New Caledonia as a region quite unexplored. Everyone on board, in fact, could not avoid hoping for frequent landfalls. Because the General could not know when it would be possible to secure fresh supplies of water, and as they would soon be entering the torrid zone, he had taken the necessary precaution to limit everyone to one bottle of water a day.

That precaution, annoying as it was for men on a salty diet and whose wine supply was in a bad state, proved to be wise. The great reefs that protect New Caledonia were known to exist, but they remained uncharted. Approaching the island from the south, the ships had to proceed northward between the Ile des Pins and perilous rocks known as the Grand Récif Sud. In effect, the approach placed them inside the great barrier reef where the sea was tranquil; but, as the coastline was uncharted, the only prudent course was to find a passage through the reefs into open sea. On 20 June 1792, after four abortive attempts had brought them close to disaster, Lt. Rossel, the officer on duty then commanding the *Espérance*, managed to find a passage on his fifth attempt, and the ships passed beyond the dangerous coast.

Thereafter, they sailed northward parallel to the coast, seeking a breach in the barrier reef of sufficient size to enable them to anchor inshore behind the shelter of the reefs. On the 24th, they finally saw a promising opening, but the winds were too strong to permit launching a boat to take soundings. On the 28th, they had reached the northern end of the island without making a landing. The lookouts could see, from the tops of the masts, that the chain of reefs they had been following extended on to the northwest. The direction of the winds prevented them from turning to the southeast to reach the east side of New Caledonia. All they could do was to continue to chart the reefs, which remain known to us today as the Récif des Français and the Récifs d'Entrecasteaux. That reconnaissance was completed on 2 July 1792, and they took aim at Cape Saint George the following day.[187]

Even though Entrecasteaux's goal was to pass through the Saint George Strait to reach the Admiralty Islands, locale of the alleged sighting of French naval debris, he was anxious to follow the west coast of Bougainville Island in the Solomons en route. Early on the morning of 10 July 1792, the Treasury Islands became visible, but no attempt to land was made. While they were approaching the southern end of Bougainville, a drenching rain brought dense fog to the island, obliging them to wait until the 13th to begin reconnaissance of its west coast.

Only as they sailed northward did the forested island seem increasingly inhabited. Fine plantations of coconut trees along the shore indicated a substantial population. At noon on the 14th, they noted a large bay that might provide a suitable anchorage, until the appearance of shoals and a sandbank covering the entrance defeated that hope. By the end of the day they had reached the channel separating Bougainville from Buka Island.

During the night, water currents carried them to the north side of Buka, and, on 15 July 1792, large coconut plantations were again a sign of substantial population. Natives came out in both pirogues and catamarans, but were at first hesitant to approach close enough to engage in trade. Their reserve was only broken when pieces of red cloth, handkerchiefs, and empty bottles were thrown toward them. They seemed to be aware of trading customs, taking care to indicate the price they put on their goods. A very fine bow was obtained, along with some arrows, in exchange for handkerchiefs that were passed to them on a cord. As they did not see any such weapons among the French, they demonstrated the use of the bow to make them understand its value and price.

"One of our cannoneers," Labillardière noted, "went to get his violin to play several different tunes. Then one of the officers, Lt. de Saint-Aignan, an accomplished violinist, played a fast-paced melody with double-stopping. At first the natives listened very attentively, astonishment showing on all faces. Soon they could not contain their joy. Various arm movements accompanied the beat perfectly and, with the agitations of the entire body, were unequivocal signs of their sensibility, which greatly pleased us. They offered many items in exchange for the instrument, but it was generally understood on board that it must not be given up. It was the only fiddle we had that enabled the crewmen to dance. With so many months of campaign still ahead, we could not deprive ourselves of an instrument that promoted an exercise so useful for the health of seamen.

187. Rossel. 1808. *Voyage d'Entrecasteaux*. 1: 103-115; Labillardière. 1800. *Relation*. 1: 192-212.

"Alas, what had begun as fair trade soon deteriorated into deceit, perhaps because too many presents had also been heaped upon the aborigines. Soon they began to refuse to release merchandise they had promised in exchange for goods received. It appeared to be bad faith but was perhaps only misunderstanding." From Labillardière's description of these natives, they were probably of Papuan stock. That is, skin dark but not deeply so, of middling height and very muscular, very thick woolly hair, their bodily hair mostly plucked; and their language was not Malay.[188]

Early on 16 July 1792, they sailed on toward Cape Saint George; that is, the southern tip of New Ireland Island. Their first anchorage, on the 17th, was close in to the offshore Ile des Cocos, the water being deep. They went on shore expecting to find many coconut trees, but barely obtained even a dozen of the nuts. This anchorage was within what they called Carteret Harbor, on the southwest coast of New Ireland. While an opportunity to cut firewood and draw fresh water, the work had to be carried on under continuous torrential rains.

The flora had mostly been described by earlier authors, but Labillardière took delight in observing plants he had never before seen: *Barringtonia butonica* J.R. & G. Forst., a mangrove whose branches extended out over the seawater; various species of fig trees from whose tops numerous aerial roots hung downward into the ground, bringing forth new stalks. The fruits of various species of *Pandanus* (screw-pine), *Heritiera*, and *Barringtonia* were floating along the coastline. The sago-palm, *Cycas circinalia* L., was abundant in the hollows, shaded by taller trees. Their drupes, if eaten uncooked, are a powerful emetic, as was demonstrated by several of the crewmen. The aborigines had learned to grill them.

The fine teak tree, *Tectona grandis* L.f., so valuable for shipbuilding, grew widely around Carteret Harbor. He found the nutmeg tree that Rumphius had described as *Myristica mas*. Its fruit was more elongate than in the cultivated species, *Myristica fragrans* Houtt.[189] Nature had also given the natives of New Ireland the species of pepper known as *Piper cubeba* L.f., the cubeb or Java pepper, whose crushed berries were valued for medicinal and culinary uses. It was widespread, a climbing shrub decorating the lower portions of tall trees.[190]

Entrecasteaux had anticipated remaining in Carteret Harbor for two weeks, but the heavy rain induced him to leave after a week, 24 July 1792, as the need for water and wood had been satisfied. A favorable wind pushed them rapidly through the Saint George Channel toward the Admiralty Islands where debris of

188. Labillardière. 1800. *Relation*. 1: 215-228.
189. G. E. Rumph. 1741-1755. *Herbarium amboinense*. 2, tab. 5.
190. Labillardière. 1800. *Relation*. 1: 230-241.

the La Pérouse expedition had been reported. Unfortunately, there were a great many islands in that archipelago; and, on 29 July, Captain Huon of the *Espérance* came on board the *Recherche* to confer with the general about coordinating the search. As the information they had received in Cape Town had seemed to refer to the more easterly area of the island chain, they agreed to reconnoiter the east part first.

At that moment they were anchored off Jesus Maria Island. The few natives they could see were quite dark and disappointingly nude. It was agreed they would proceed to a small island to the east-northeast that could be seen, Vendola Island. There they came to a halt about 1:30 p.m. An armed longboat carrying items for trade was launched from each ship. They reached within about 100m from shore where protective reefs prevented any further advance. The frigates stood within range to protect the boats from any attack during trading on the reefs. It was known that, when Captain Philip Carteret had reconnoitered the southern Admiralty archipelago in 1767, aborigines had shot arrows at his people on several occasions despite lavish displays of friendship. These natives made elaborate gestures to entice the crews to come ashore, but that knowledge, plus the reefs, dictated caution.

Vendola Island appeared to be cultivated to its summit. Some of the plots were fenced, suggesting a recognition of property rights. Fine coconut plantations were evident around the base of the mountain, and the more elevated places appeared to be reserved for the cultivation of various root crops used for food. The natives seemed to be relatively indifferent to most items offered in trade, but wanted iron and were delighted to receive iron nails.

Several of the men, wearing double rows of small seashells around the forehead, enjoyed considerable authority and were evidently accustomed to obedience. Each one carried a staff and made himself noticeable by the blows he distributed to those to whom he gave orders. Labillardière remarked that the French had not expected to see a man so treated amongst a people seemingly so close to the state of Nature. Yet, all the natives seemed to have happy faces.

But, alas, there was no evidence of anything of European origin: neither clothing nor arms. The natives did wear ornaments of white shell and belts of a dull red. These colors, against a dark skin, could have been perceived at a distance as French naval uniforms, especially if the viewer had been predisposed to see such evidence. Nothing to do, therefore, but to proceed to Manus Island, the only large island in the Admiralty group.[191]

191. Ibid. 1: 241-255; Rossel. 1808. *Voyage d'Entrecasteaux*. 1: 130-136.

On 31 July 1792, the frigates moved close to shore. The natives in pirogues made friendly gestures, obviously wanting the French to land. The offer of nails and knives evoked shouts of pleasure, as these natives, too, knew the use of iron. A vigorous trade ensued in utter good faith with no intent to steal, in contrast with what the French had just experienced on Vendola Island, where the natives had resorted to theft once they recognized they could steal with impunity. On Manus, the natives made every effort to fulfill obligations.

Another unique custom on Manus was the use of the the seashell from the genus *Bulla*. One of them was attached to the end of each man's penis. An opening had been made above the enlarged part of the seashell to enable it to fit over the head of the penis. The ornament provided the French sailors much merriment, and its whiteness was a striking contrast to the dark skin. It proved to be difficult to induce any native to part with his ornament in trade; but when accomplished, its removal revealed that these natives, unlike most of the inhabitants of the South Seas, did not practice circumcision. The pressure from the seashell often produced a white tumor on the upper foreskin. The only male child they saw was not wearing such a shell, indicating its adoption depended upon reaching a certain age.

Labillardière was struck, above all else, by the unique character of the aborigines on Manus, notably different from the inhabitants of nearby Vendola. The two peoples possessed essentially the same crafts, yet the Manusians seemed to be extremely gentle, very peaceful, and had given evidence of great honesty. Could it be he wondered, as the two groups had comparable technical advancement, that the islanders of Manus had had better behavior imposed upon them? That is, thanks to inhabiting a much larger island than Vendola, they had been more frequently visited by larger (European) ships over the past several centuries. Could they, in time, have come to learn that successful trade depended upon obvious integrity or good faith? Speculation, to be sure, but a reflection of that 18th-century mind, which assumed that progress required civility.

Labillardière's favorable view of the Manusians did not blind him to a troubling survival of regional tradition. The chief who commanded each pirogue that came out to trade usually required that all items given to the paddlers be surrendered to him and was ready to use violence to enforce the seizure. One of the paddlers, who had received a small piece of red cotton serge, prized it so highly that he did not give it up to his chief, bringing upon himself a great number of blows from a staff. Another paddler received the same treatment when, preoccupied by the proximity of the frigate, he was neglecting to bail out the water that was filling the pirogue. An undercurrent of conventional brutality disturbed any onlooker hoping to see sweetness in a people so close to Nature.

Several native products were offered for trade by the chiefs alone, products apparently reserved for their use: Gourds filled with a lime that had been reduced to a fine powder, and packets of leaves from the Indian pepper plant, *Piper siriboa* L. The powdered lime, if held in the mouth, presumably produced a very pleasant sensation. The pepper leaves were to be chewed, and evidently without the areca nut, as no traces of the betel were evident on their mouths. These were among the sensual privileges of Manusian aristocracy from all appearances.

Unfortunately, the French saw nothing in the possession of these islanders that could have belonged to Europeans. Before midday, they raised considerable canvas to sail along the coast in search of a safe harbor for refreshment. That failing, Entrecasteaux necessarily turned westward toward New Guinea through uncharted waters. On 2 August 1792, they passed through the Hermit Islands; but between 4-12 August their progress was maddingly slow because of contrary winds. Their goal had now become the island of Ambon, the seat of Dutch government over the various Dutch possessions in the Moluccas.

The several weeks before their arrival in Ambon became a torture. The equatorial heat off the New Guinea coast accelerated the decomposition of the water supply taken on at Carteret Harbor, and refreshment was urgently needed. Scurvy, having already made its appearance among the crew, was making alarming inroads by late August. Nearly all the afflicted experienced very sharp pains in the loins. One of the first symptoms of scurvy was the appearance of whitish tubercles, often the size of a chicken egg, on different parts of the body, initially especially on the arms. Labillardière noted that the victims' skin was not covered by what were called scorbutic spots. In his opinion, scurvy invades the cellular tissue of a lymphatic humor very rapidly, hardly altering skin color.

He also accepted the common notion that the use of salted meats was one of the principal causes of scurvy, yet was restrained from certainty by the presence of two victims on board known to have refused to eat salted pork. Given his earlier remarks about antiscorbutics, and his recognition of the importance of having fresh fruits and vegetables on board, it is unclear whether he connected the latter specifically with the former, especially as he made no mention of James Lind's important *A Treatise on the Scurvy* (1753) that had identified lemon juice and fresh fruit as antiscorbutics.

The passage south through the Moluccas was continually inhibited by reefs and sandbars, not to speak of capricious winds. Only on the evening of 5 September 1792 did they reach the island of Ambon. The next morning an officer was sent ashore to obtain permission from the governor to anchor in his port. Given the evident need for substantial rest, the general's intention was to remain for at least a month.[192]

On 8 September 1792, the naturalists asked Entrecasteaux to present them to the Dutch governor, as the General had overlooked them when taking all the ships' officers to meet the governor on the 6th, a slight that Labillardière chose to record as a simple oversight. The naturalists were anxious to have the local Dutch be aware that they were on a scientific mission so that all facilities to pursue their research would be made available. They were given the same permission to reside in town that had previously been granted to the naval officers and rented a house for the duration.

This required transferring the equipment the naturalists would need for the preparation of their collections from ship to shore. When they arrived with their effects at the rented house, they found it already occupied by officers from the two ships, who knew quite well that the naturalists had rented the place. They had somehow induced the landlord to give them the key. The ruse amused the officers very much. Even though it proved to be easy to find another lodging, the incident only added to the treasury of grievances. "We had not believed them capable of such mischief," Labillardière commented.[193]

The naturalists had been sufficiently debilitated by the recent weeks at sea that they limited their outings for the next several days to the town's immediate surroundings. On 15 September, they departed very early in the morning, but the heat became so enervating that they retreated to their house by noon. At that point, Louis Ventenat, the chaplain-naturalist from the *Recherche*, became so critically ill that the others could not leave him for the next four days. His symptoms were frightening: frequent vomiting, extremely foul evacuations, a trembling in the tendons, feeble pulse, and exhaustion. Violent spasms within the lower extremities were very painful.

Although they suspected the fever was highly contagious, they refused to leave their traveling companion as long as he remained in danger. His symptoms became even more alarming on the 16th: frequent bouts of hiccups, the pulse increasingly irregular, and features convulsed. But near daybreak, after a frightful night, the symptoms gradually abated, the harbinger of an abundant sweat that could draw their friend back from the gates of death.

Meanwhile, the treasurer of the Dutch company, Mijnheer Bourguelles, had convinced Entrecasteaux that the collective knowledge of all the physicians of Europe was worth less in the treatment of such an illness [probably a dengue fever]

192. Rossel. 1808. *Voyage d'Entrecasteaux*. 1: 136-154; Labillardière. 1800. *Relation*. 1: 256-288.
193. Labillardière. 1800. *Relation*. 1: 294.

than the science of a Malay doctor. Accordingly, one of the most accomplished of them had been summoned. The French physician-naturalists allowed him to proceed with his alternative treatment so long as he should cause no harm.

To their surprise, the Malay doctor did not bring any herbal remedies to be taken internally. Instead, he rubbed the patient's skin, especially massaging the lower extremities. Next, he pronounced, in mysterious tones, words seemingly addressed to a Supreme Being and, as was explained to the French onlookers, then exorcised the evil spirits that the islanders believed to be the cause of illnesses. Finally, he prepared to inundate the patient with a bucket of cold water freshly drawn from a well. At that point the French doctors called a halt, fearful that such treatment could suppress the critical perspiration they had been anticipating because of the improvement in the pulse. The patient's revival conformed to their expectations.

By 19 September 1792, Ventenat was well enough not to require such assiduous care, and his colleagues could take to the field. Following a stream that flowed into the harbor from the west, they acquired a great load of fine plants. On the following day, they walked more southward, coming across a stand of handsome palm trees. The natives called them *sagouer*, and Rumphius had described them as *Sagurus* in his herbarium.[194]

The peduncles of their inflorescence, when freshly cut, secreted a very pleasant liquor, which, in that hot climate, could ferment very quickly and become acid had the natives not learned to store it in containers made from the wood of *Soulamea*. For two months of the year, a palm could produce 6-8 litres of liquor every day, providing care was taken to reopen the incision on the peduncle every day. The liquor obtained especially during the heat of the day contained sugary bits that the Malays called *goula itan* (black sugar), close to the color of chocolate, which they used in preference to any other form of sugar. It was much cheaper than cane sugar. Labillardière named the tree *Arenga saccharifera* Labill., a new genus as well as a new species.[195]

When working on his recent collections on 23 September 1792,[196] Labillardière recognized preservational problems he had not previously encountered. It had been logical to assume that the equatorial heat of the Moluccas would promote

194. Rumphius. 1741-1755. *Herbarium amboinense*. 1: tab. 13.

195. Labillardière. 1800. *Relation*. 1: 301-302.

196. When Labillardière was preparing his journal for publication in 1800, the Republican calendar of the French Revolution was still in effect. Consequently, as a good Republican, he converted all the Gregorian dates in his journal, beginning in September of 1792, to their Republican equivalents. This is confusing for the reason that the Republican calendar did not go into effect until September *1793*. Wherever possible, I have tried to reconcile his dates with those of Entrecasteaux in Gregorian terms, otherwise necessarily approximating them as there was no Republican calendar for the next 12 months.

quick dessication of specimens, but the contrary was proving to be true. The air was apparently deriving so much humidity from the sea that the climate, in fact, retarded dessication and threatened to ruin specimens unless special attention was given to the collection as quickly as possible.

Labillardière and Riche, at least, were no more deterred by the difficulty in preserving collections than they were discouraged by the knowledge that many of the plants they would see on Ambon had already been collected or described by G. E. Rumph, Jan Burman, or C. P. Thunberg. But it becomes noticeable that we never encounter the name Louis-August Deschamps on their outings, giving credence to Ventenat's snide observation that Deschamps never ventured far from breakfast, lunch, and dinner.

Early on the morning of the 24th, Labillardière and Riche turned eastward along a pretty stream. Several species of *Ludwigia* L. on the surface of the water were noted with admiration, in particularly how Nature had provided for the preservation of the plants. Large tubercles, oval in form and filled with air [aerenchyma], had been placed along the stalk to enable a plant to float on the surface.

Elaeocarpus serrata L. was another wonder; it was covered down to its lower branches with beautiful flowers. Despite the density of the foliage, which the sun could hardly penetrate, parasitic orchids, attached to the largest of tree trunks, had an astonishing brilliance. In places less wooded, they were delighted to see *Cussonia thyrsifolia* Thunb. in the Araliaceae, whose large palmate leaves were strikingly ornamental. An abundant harvest of specimens only enhanced the joy of the outing, and Labillardière was forced to devote the entirety of September 25th to the preservation of plants.

The subsequent days on Ambon were similarly occupied, although on several occasions the botanists would engage outriggers that could take them some kilometers along the coast to permit them to penetrate a new region. What is strikingly absent from Labillardière's journal, as well as from Entrecasteaux's, is any reference to a scandal that is said to have surfaced during the sojourn on Ambon. We are told that Captain Huon of the *Espérance* had been informed that Labilladière and Riche were conspiring to produce their own narrative of the expedition, for which the nautical information would be provided by three naval officers, Willaumez, Legrand, and Laignel; a narrative that would be strongly Republican in bias and would be published on Mauritius before the return of the expedition to France. The report further indicated that Entrecasteaux had been shocked by the news, as surely he would have been, as the private publication of a journal would have been a violation of the king's instructions, and because he believed that the naval officers accused were entirely loyal to him. Consequently, the report concluded,

Entrecasteaux had reprimanded the naval officers, and their collaboration had presumably ceased.[197]

The story was based upon a letter written by a sub-lieutenant on the *Espérance*, Jacques-Malo la Motte du Portail, whose later naval career would prove the profundity of his royalist convictions. At the time of the expedition, moreover, he was a malcontent, believing himself to have been inadequately promoted and resentful of fellow officers whom he believed had been inappropriately advanced. In the absence of any additional documents to provide evidence of a conspiracy, the reliability of Portail as a witness is questionable. Had he embroidered the undeniable shipboard irritations into a self-serving scandal? Was his version of events, promoted by shipboard gossip, responsible for the ultimate segregation of the five he named as disloyal Republicans at the end of the voyage? Or had there been, in fact, a collaboration in defiance of the king's orders?

Labillardière, obviously, would publish his journal after the expedition and would acknowledge Ensign Legrand's nautical information in the preface. Whether Labillardière maintained his journal in the expectation of eventual publication is unknown. His later efforts to conceal his journal to avoid confiscation could be attributed to an intention to publish an unofficial narrative of the voyage. But he had also confided to that journal important geographical information about his plant collections, field notes if you will, as well as precise diagnoses of new genera and species, references essential to the eventual publication of his plant collections.

There remains the matter of reliable identification of political commitment. Labillardière was certainly Republican, then and thereafter. Sub-lieutenant Laignel had been a member of the Society of Friends of the Constitution of Brest, which had promoted indiscipline within the naval forces based there. But surely Captain Huot would not have engaged him for service unless confident of his loyalty. Riche had repeatedly annoyed Entrecasteaux with complaints about inadequate assistance in obtaining zoological specimens. Was that proof of republicanism? or simply professional ambition? Of Legrand's and Willaumez's politics, we can be even less sure. Would Willaumez have ever been promoted to be vice-admiral (1819) under Louis XVIII had anyone seriously believed him to be of questionable loyalty to the crown? All questions unanswerable and made more intractable by the failure of both Rossel and Labillardière, who had seen Portail's accusations when they were editing their own works, to comment.

197. Horner. 1995. *Looking for La Pérouse*, p. 106.

Instead, we have an entire chapter that Entrecasteaux devoted to observations on the governance of the Dutch East India Company, the productivity of Ambon, the mixed population, and its religions.[198] While Labillardière was narrating his botanical excursions and describing the difficulties of plant preservation. It becomes possible to suspect that not everyone else on board those ships had occupations sufficiently demanding to confine them to useful knowledge.

On 28 September 1792, Labillardière sailed at least 10k east of the town in a double-outrigger pirogue to reach a more mountainous terrain and different soils. *Canarium commune* L., growing amidst limestone rock, was the tallest of the local trees, a species in the Bursaraceae notable for producing an aromatic gum or resin. He found many stalks of *Cynometra cauliflora* L. growing in the woods, a legume also cultivated by the Malays as *nam-nam*. The taste of its fruits was similar to a slightly acid apple. Work on his collections occupied him for the entirety of the next day.

On the 30th, the outrigger was taken about 6k across the harbor in the opposite direction. Immediately on shore, he encountered shrubs in the Goodenia Family named *Scaevola lobelia* Murr., which he had previously seen on New Ireland. At the high tide mark, he found trees of the genus *Heritiera* Dryand., reminding him of his botanical friend in Paris. He also found the handsome laurel tree now known as *Cinnamomum culiliwan* Blume that produces, by distillation, a very prized aromatic oil.

The tallest of the clove trees seen on that outing, *Syzygium aromaticum* (L.) Merr., were no more than 7m tall, their trunks about 2dm thick. The natives took care not to expose the cloves to the rays of the sun to preserve the essential oil, and the visitors saw cloves spread out to dry on mats under covered sheds. Labillardière commented that the natives were obligated to turn over their aromatic produce to the Dutch company for about 1/150 the price it would be sold for in Europe.

On 2 October 1792, the naturalists were invited to examine the governor's natural history cabinet. Its outstanding feature was a large collection of beautiful butterflies, perfectly preserved; but there was also a fine selection of seashells. They were given to understand that the taste for collecting natural history specimens was widespread within the Dutch community; and a judicious distribution of rare specimens to people of credit, in either Batavia or Europe, had become a standard ploy for self-promotion.

The next day, they took their usual pirogue toward the sandy shore near the harbor entrance. Claude Riche had the native paddlers, described as Papuans, dive

198. Rossel. 1808. *Voyage d'Entrecasteaux*. 1: 155-166.

for valuable marine specimens while Labillardière went ashore. In marshy terrain, featuring a great number of species, Labillardière noted *Acanthus ilicifolius* L. with its holly-like leaves, as well as a variety with its leaves entire. Reembarking, they followed the coastline closely and could see the brilliant-red flowers of *Erythrina corallodendron* L., a legume. The screw-pine, *Pandanus odoratissima* L.f., could be seen on a slope of gray rock, its branches bent toward the sea. Farther on, in a plantation, they met *Averrhoa carambola* L. whose fruit resembled gooseberries in taste and had a pleasant acidity; as well as *Sesbania grandiflora* (L.) Poir. with its edible legume.

For the next few days, Labillardière felt obliged to work on his collections, which required diligent care if they were to be preserved in that climate. He complained that an intelligent assistant would have spared each naturalist valuable time that could have been used in a more profitable manner. The moment of departure from Ambon was approaching, and specimens had to be ready to be packed aboard. This may be why, during the final days, he gave increasing time to recording observations of the local population rather than collecting additional material. The ships had been caulked; and the rigging, having been scrupulously inspected, had been found to be in good condition.[199]

Over the prior centuries, the Molucca Islands had long been dominated by Arabs (the Moors of Ceylon) and by the Malays. Only later had they been dominated by Europeans: the Portuguese, the Spanish, until the Dutch had finally become predominant. Labillardière quite understood that each successive sovereign had brought such changes to the customs of the natives of Ambon that it had become nearly impossible by the end of the 18th century to detect what their aboriginal character had been. The natives had lost their native character except for their darker skins.

The successive waves of Islam, Catholicism, and Protestantism were a reflection of the various foreign dominations. Labillardière perceived that the schools, established for the instruction of the native population, emphasized the Malay language and Protestantism, omitting instruction in Dutch entirely: for that was the language of politics and commerce. The Chinese had appeared in the Moluccas as merchant-traders. The Dutch company had only allowed ships coming from China to trade through the ports of Macassar (Celebes) and Batavia (Java). A Chinese trader who wanted to specialize in inter-island trade had to become naturalized and promise never again to return to China. No other foreigners were permitted to reside in the Dutch East Indies. Such regulations enabled the Dutch company to concentrate on, and monopolize, the trade with Europe.

199. Labillardière. 1800. *Relation* 1: 302-345.

Labillardière tried to put the slave trade, a trade he profoundly abhorred, into a larger context. The local slaves seemed largely to have been drawn from Macassar or the nearby island of Ceram. Before the arrival of the Dutch, the islanders of Ceram had customarily eaten prisoners taken in battle. It was painful to learn that they had only abandoned that atrocious custom when they had found they could derive an even greater profit from the sale of their captives to the Dutch. If that development would seem to have been an improvement, it had not necessarily been so, as it became the motive for more intensive tribal warfare. Even so, Labillardière concluded, if the introduction of slavery had to be reckoned as an upward step for mankind toward civilization, then the prior state of mankind had been extreme degradation. His conclusion did not square comfortably with the fashionable faith in the natural goodness of man.

The Dutch rationale that made slavery a humane advance was hardly less comforting. Labillardière recalled an encounter, as he was returning from the field on 26 September 1792, with a slave whose decrepitude aroused his curiosity. With the aid of the native guides who had accompanied the naturalists, he tried his best to find out the old slave's age, but in vain as he apparently knew nothing about it. The naturalists also came to realize that he also did not know how many years he had passed in servitude: no concept of time, no past.

Besides securing a good supply of fresh water in preparing to leave Ambon, a buck and two does were taken on board the *Recherche* in the hope of enriching Australia with a fine species. They also took on a good supply of chickens, ducks, and guinea fowl; colorful bunches of bananas and gourds; sweet potatoes and yams; pigs and goats. The ships's butcher was in charge of feeding the livestock. Unable to secure a satisfactory forage grass, he had bought thick stalks of banana trees that provided the animals with a succulent food, its moisture a welcome supplement to the limited ration of water alotted to them.

In the markets of Ambon, bamboo shoots were sold as vegetables. An ample supply was acquired as, sliced and preserved in vinegar, they could provide an excellent food for a long voyage. A stock of sago palm proved to be much too large, as the crewmen never accustomed themselves to it. No matter the rational arguments about its healthful properties, they developed such a dislike for it that they preferred the worst of the salted meats.

No wine had been available on Ambon to supplement their few remaining casks of drinkable wine. They bought several barrels of arrack, the Asian beverage similar to rum, but found it not the equal to even a mediocre wine eau-de-vie. No other alcoholic liquor could be obtained.

We learn for the first time that the Recherche had had a milk-cow on board. By the time the ship reached Ambon, her milk had become tainted. It had been hoped she could be exchanged for a fresher animal, but cows proved to be unavailable. The domesticated buffalo common in Asia was an alternative, but the female gives only a little milk. Beyond which the animal is nearly unmanageable, can be dangerous, and is quite unsuited to life on board ship.

At 7:30 on the morning of 13 October 1792, a favorable breeze allowed them to weigh anchor. They proceeded down the channel, passing the end of the harbor about 11 o'clock, setting a course for the Australian coast.[200]

200. Ibid. 1: 354-367.

X
The Voyage of Entrecasteaux: Part 4
Esperance Bay

Consider for a moment the charms of a cruise on a small ship in the tropics with a goodly supply of livestock and barnyard fowl between decks. In addition to this supply, the General had permitted crewmen to bring pigs and fowl aboard for personal use. Labillardière felt obliged to make note of the *congestion* and the *sickening odor* emitted. Rarely have the personal sacrifices made for the advancement of science been so poignantly depicted.

On the second day out, 14 October 1792, a Dutch cabin boy emerged from hiding on the *Recherche*, at about the moment the captain of the *Esperance* signaled that he had just found six fugitives from Ambon on board: three soldiers deserting from the Dutch company's garrison, one sailor, and two Black slaves. All were fleeing from a place where they had suffered nearly equally in servitude. Their presence worsened the glut of beings on board, but no one thought to turn back to land them.

They sighted Timor Island on 19 October 1792 and proceeded along its west coast toward the Sawu Islands, which they sighted on October 25th. Entrecasteaux was anxious to reach the Australian coast and, thus, finally regain the projected route of La Pérouse. They soon lost all sight of land and would not see land again until they reached southwest Australia. Entrecasteaux meant to go ashore in Dirk Hartog Bay (later known as Shark Bay), but contrary winds led him to abandon that plan on 14 November 1792.

The livestock, meanwhile, were not all living up to expectations. Numerous parrots, bought by many crewmen for entertainment, did not readily adapt to life on a ship. Their piercing cries gave no one a moment of tranquility during the day, and a few of them died virtually every day, evidently not to universal regret. The mortality among the chickens was also high. Labillardière blamed the butcher for not covering their cages at night with a canvas.

Within two weeks of departure from Ambon, he was complaining about the water taken on at Ambon, attributing its foul character to inadequate cleaning of casks before refilling. On 16 November 1792, someone left too large an opening in the waist, permitting their one buck to fall into the sea. Captain Huon was notified of the accident and urged to take precautions with the buck he had on board. But it died before they reached a landing place in Australia. One of the does on the *Recherche* had been fading fast and had to be surrendered to the butcher as she would not live much longer. By the 21st of November, Labillardière added that the goats were perishing from day to day, apparently for want of suitable nourishment.

Two days later, they rounded Cape Leeuwin, now the site of Augusta, but the wind was too violent to permit approaching shore. The next catastrophe was the death of the ship's blacksmith, found dead in his bunk of a stroke after over-indulgence the previous evening during a small festival organized by the gunners. The loss of such a critical craftsman would have been irreparable were it not for the presence of an intelligent German mechanic who had come aboard at Cape Town as a stowaway.[201]

On 6 December 1792, along the south coast of Australia, they perceived a large bay, apparently very sheltered, and protected by a long chain of outer reefs. Entering the bay was clearly going to be difficult and risky, but the urgent need for fresh water meant that a route had to be found. Finally, on 9 December 1792, the *Espérance*, guided by Ensign Legrand, discovered a suitable passage into the protected water. Soundings were taken from a small boat, finding an excellent bottom at a depth of 15m and only about 300m from shore. After the successful anchorage, the *Recherche* joined her in the harbor. The bay was initially named for Legrand but was subsequently called Esperance Bay.

On December 11th, Labillardière went ashore on an islet just to the southwest of their anchorage. About 4k long, 2k wide, and hilly, it has recently been identified as Woody Island, close to the east side of Esperance Bay.[202] The plant-bearing soil, limited to the least rocky places, was covered with shrubs, often densely. He made an abundant harvest of plants, several of which he believed to be new: a magnificent *Leptospermum* with silvery leaves and brilliant red flowers (later transferred to *Agonis*);[203] two new species of *Pimelea* in the Tymelaeaceae;[204] plus a new *Lobelia*.[205]

He had to spend the better part of the following day describing and preparing his collection. Meanwhile, a suspicion was developing on board that water might not be found on the arid coast, and no sign of human habitation had been seen. Accordingly, Captain d'Auribeau ordered everyone, except the General and himself, to be limited to three-quarters of a bottle of water a day. Labillardière thought poorly of their exclusion from the general rule.

201. Ibid. 1: 368-381; Rossel. 1808. *Voyage d'Entrecasteaux*. 1: 167-175.

202. J. H. Willis. 1953. "Plants of the Recherche Archipelago, W.A." *Australian Geographical Society Reports*. 1 (3a): 3-6. Willis, a botanist, did not realize the Labillardière's dates during the "Year I" were unreliable.

203. *Leptospermum marginatum* Labill. 1806. Nov. Holl. Pl. 2: 10. Tab. 148. SY=*Agonis marginata* (Labill.) Schau.

204. *Pimelea spathulata* Labill. 1804. Ibid. 1: 9. Tab. 4; and *Pimelea drupacea* Labill. 1804. Ibid. 1: 10. Tab. 7.

205. *Lobelia heterophylla* Labill. 1804. Ibid. 1: 52. Tab. 74.

He became further insensed upon learning that Beautemps-Beaupré, the hydrographer, had been dispatched in the longboat early on the morning of 13 December 1792 to reconnoiter the islets in the small archipelago, both to find a passage out to the east and to seek a suitable watering place. Although he recognized the distinction between hydrographic operations and the naturalists' research, Labillardière had desired to take part in such a venture and believed that the arrangements for it had been made *clandestinely* to omit him. As the longboat was not available to convey him to the mainland, he had to settle for a second day on Woody Island.

By following the contours, he reached one of the higher points on the southwest part of the island. To his great joy, he found a feeble current of fresh water emerging from fissures in granitic rock, so small it took some time to supply the needs of his small party. On another hilltop, limestone had been deposited in nearly horizontal layers, an alteration in terrain that produced plants he had not previously seen. One was a new eucalyptus, whose name was derived from the form of the calyx, very elongate, pressed outward by the stamens as they develop, deciduous after reaching full growth.[206]

The members of the party resolved to spend the night on Woody Island, possibly influenced by the pleasure of drinking their fresh water without restriction. Labillardière had shot birds for his collection that day, but gave some penguins to his companions for grilling. They found a grotto in the rocks that provided shelter from the wind, built a fire, and expected to spend a comfortable night. It was not to be. A sudden violent storm, reversing the wind direction and bringing rain, forced them out of their grotto and extinguished the fire. The half-cooked penguins proved to be very good, but everyone regretted not having returned to the ship.

With the light of day on 14 December 1792, he collected many plants in the marly soil at the base of the hills toward the southwest. There he found one leguminous plant with a papilionaceaous corolla and distinct filaments upon which to base a new genus, *Chorizema*: leaves simple, alternate, sessile, long, dentate, coriacious, with 2 spines for stipules. The form of the leaves [suggesting those in Ilicaceae, now Aquifoliaceae] accounts for the name of the type-species.[207]

On that very morning, when the mapping of the islands and reefs was continuing in what would ever after be known as the Archipelago of the Recherche, Captain Huon sent a boat ashore from the *Espérance* bearing several officers to make astronomical observations and Claude Riche, the naturalist. Huon gave

206. *Eucalyptus cornuta* Labill. 1800. *Rel. Voy. Rech. La Pérouse.* 1: 403. Tab. 20.
207. *Chorizema ilicifolia* Labill. Ibid. 1: 403–404. Tab. 21.

them instructions to return by early afternoon. The three published accounts of what occurred thereafter on shore vary in detail, but they are consistent in describing the anxious 54 hours during which Riche had become lost. In the end, after Riche's reappearance, those anxieties gave way to complaints and recriminations, which would exacerbate extant irritations and hostilities beyond repair.

Once on shore, Riche had gone his own way to collect specimens. The rest of the party set out to investigate the source of fires whose smoke had appeared from place to place. No natives were seen, at least on that day, giving rise to the supposition that the fires had been set to conceal the direction of their flight. With the exception of Riche, the party gathered at the boat at the appointed hour. When he had not appeared by 5 o'clock, there was reason for alarm. The country was a desert, it was extremely hot; and Riche, fair-complexioned, was vulnerable to the sun.

By 7 o'clock, Lt. Dominique de La Grandière, commanding the boat, had judged they could wait no longer. The boat was anchored in a place that would become dangerous should the weather change and the sea rise, and they had little remaining food. Along with a coat, weapons, hardtack, and eau-de-vie, he left a note saying they would return at dawn the following day, weather permitting.

As the weather was favorable, a boat was again sent on 15 December 1792 to search for Riche. The items left for him were found undisturbed, so the search party proceeded in the direction they had seen him take the day before. During that day some aborigines were perceived, but they fled every time they were approached, preventing any dialogue. When the party returned to the *Espérance* that afternoon without Riche, Captain Huon went at once to the *Recherche* to determine with Entrecasteaux what measures should be taken in that sorry situation. Entrecasteaux invited Labillardière and Deschamps to join the discussion.

It was evident to all that Riche, by going alone and without supplies into alien land, could not long survive on its burning sands. And, in view of the presence of aborigines, Entrecasteaux feared the worst. What is more, the dwindling water supplies on the ships, unable to take on fresh water at that anchorage, amounted to an urgent need not to prolong the anchorage. Captain Huon argued that remaining any longer would put the expedition in jeopardy. Given the little hope of finding the naturalist alive, he emphasized the disadvantages of remaining where they were. Deschamps, convinced by that argument, was the first to express a vote for moving on.

To counter that vote, Labillardière recalled that in December of 1777 two of Captain Cook's sailors had gone astray on Christmas Island. Cook had sent out several parties to conduct a careful search. One man was recovered after a day, the other after 48 hours. Surely they should do no less for their unfortunate friend.

Entrecasteaux concurred. He had two search parties formed and provisioned for 4 days, putting Labillardière in one party and Deschamps in the other. They went ashore late that very afternoon. The General ordered a cannon to be fired every half-hour so that Riche, if still alive, would know the direction of the anchorage.

The parties went inland in two directions but soon returned to their landing place because of nightfall. During those moments Labillardière had been struck by a beautiful plant growing in the arid ground that had affinity with *Iris*, but was also close to the genera *Dilatris* and *Lanaria*. Yet, because of its irregular corolla, he thought it represented a new genus, giving it the name *Anigozanthos* Labill.[208]

At dawn on 16 December 1792, the two search parties again set forth, each armed with a compass. After going northward at least 10k across plains of calcereous sand, Labillardière noted a narrow bottomland that was comparatively green and suggesting fertile soil. Occasional holes with sweet water were present, too distant from the ships, alas, to be of any use. Amidst limestone rocks, he collected beautiful plants, many of which were within the Proteaceae. He cited two new species of *Banksia* that he would publish and illustrate.[209] He also found *Eucalyptus cornuta* Labill. there and many other species in the Myrtaceae.

Vigorous walking finally brought his party to the edge of a salt lake, apparently connected to the sea, and since identified as probably Lake Warden.[210] They were soon joined by members of the second search party, which also had reached the lakeshore to the northwest. Shoeprints had been found in the moist soil near the lake, no doubt Riche's, and footprints from bare feet near his. His pocket handkerchief and one of his pistols had been found a few steps away. A bit of smoke still rose from a fire-site nearby, and some scraps of paper bore Riche's handwriting.

Having thus lost all hope of recovering their unfortunate traveling companion, the searchers turned back toward their embarkation point. As they approached it, the crewman left behind to guard the boat ran up to tell them that Riche had just reached the embarkation point exhausted from hunger and fatigue, but at least alive and uneaten. He had barely survived on the fruits of the shrub now known to us as *Leucopogon richei* R.Br. in the Epacridaceae, a family with great affinity to our Ericaceae.[211]

208. *Anigozanthos rufa* Labill. Ibid. 1: 409-410. Tab. 22.

209. *Banksia repens* Labill. Ibid. 1: 411. Tab. 23. *Banksia nivea* Labill. Ibid. 1: 411-412. Tab. 24. SY= *Dryandra nivea* (Labill.) R.Br.

210. J. H. Maiden. 1910. "Records of the Earlier French botanists as regards Australian Plants." *Journal of the Proceedings of the Royal Society of New South Wales*. 44: 130-131.

211. Robert Brown would also honor Riche with the genus *Richea* in the Epacridaceae. The family has recently been transferred into Ericaceae.

Riche's own account of his ordeal indicated that he had indeed gone in several directions along the lakeshore. He had tried to approach aborigines in the hope of obtaining food; but, far from being a menace, they had retreated rather than risk a meeting. He had made an abundant collection of shells and minerals, but eventually had had to abandon them because of their weight. By turning back to what he thought was southeast, he had reached the sea. Then, turning west, he followed the beach until he glimpsed the ships in the distance and hastened forward as fast as his weakness would allow.[212]

The aftermath, which ought to have been joyful, proved to be otherwise. Labillardière, to be sure, was delighted. He had the enfeebled Riche fed only by degrees until his face, at first entirely altered, gradually reanimated. The conventional signals were given from shore to indicate that Riche had been recovered; but those on board the *Espérance* had been so convinced from the start of the uselessness of the search that they only recognized his recovery when the longboat came close enough to the frigate for him to be seen. Rather than graciously admitting they had been wrong, the officers instead began to speculate amongst themselves about Riche's motives, preferring to believe that Riche had planned to get lost. The perversity of the argument outraged Labillardière.

Entrecasteaux simply asserted that the regrettable incident, and the loss of the collections, could have been avoided had Riche been provided with an escort. Entrecasteaux also knew that Riche had not taken even his servant with him on the assumption that the excursion was to last only several hours. What followed is not fully documented: but it would appear that Riche responded, to what seems in retrospect to have been a fair criticism, by reviving old complaints that astronomy and hydrography were favored and assisted more by the General than natural history was. Riche evidently then asked for an allocation of time, resources, and personnel more appropriate to the *primacy* of natural science, a claim contrary to the instructions Entrecasteaux had received from the ministry of the marine.[213]

In any case, the charge was baseless. Entrecasteaux explained that the character of the voyage itself had made landings rarer than had been desired. Unforeseen circumstances, that is, had made frequent or lengthy layovers impossible. He did acknowledge he had given orders that the naturalists were not to embark on boats sent out to map and sound coastlines and bays. Most of that work was done offshore, and only very brief landings were authorized. Clearly the problem was logistical, not deliberate discrimination.

212. Rossel. 1808. *Voyage d'Entrecasteaux*. 1: 180-193, 196-208; Labillardière. 1800. *Relation*. 1: 404-414.
213. Richard. 1986. *Le Voyage d'Entrecasteaux*, pp. 158-160.

The episode with Riche only deepened Entrecasteaux's growing annoyance with the independence of his naturalists and their schemes. "One cannot emphasize too much," he noted in his journal, a document meant for the eventual edification of the minister of the marine, "the advantages to be gained on these kinds of expeditions by only employing people attached to the navy; because such people are better judges of what circumstances permit and never make imprudent requests, and would be less likely to attribute to your ill will those observations that have to be made to them about the impossibility of conceding what they desire."[214]

Early the next morning, 17 December 1792, the ships sailed from Esperance Bay despite the prevalence of an east wind that threatened to impede their reconnaissance of the southern coast of Australia. The need to tack, and to tack about, provided the opportunity to reconnoiter the Archipelago of the Recherche and to refine their maps but did little to advance the search for fresh water or the search for La Pérouse along that south coast. Not until 24 December 1792 did they reach the coast of what was then called Nuyts Land, which gave the appearance of being sterile, uninteresting country. Still, they endeavored to follow that coastline in the hope of sighting terrain likely to produce water.

On 2 January 1793, Captain Huon had the commanding general signaled that he had suffered some damage to his rudder. Entrecasteaux ordered an immediate halt, and Lt. Denis de Trobriand came on board the *Recherche* at once to give a fuller and, what turned out to be, alarming report. Huon, for some days, had had everyone on the *Espérance* limited to three-quarters the daily ration of water. He had been forced to eliminate antiscorbutic beverages and fresh bread for *everyone* on board. His remaining 30 casks of water, at that rate of consumption, would be barely sufficient to reach Tasmania if they sailed there directly. It was known from their prior visit that they would find water there and a sheltered haven in which to make repairs.

The expedition at that point was midway through the Great Australian Bight, off-shore just east of the present site of Eucia. To continue searching for water along a coast that seemed nearly devoid of it would be time-consuming and risky. Entrecasteaux had no alternative but to steer directly for Tasmania. That order was given on 3 January 1793. As they passed southeastward, they encountered strong westward currents, leading Entrecasteaux to wonder whether the currents might indicate a channel between Australia and Tasmania. Had he dared take the time to explore that possibility, it seems probable he would have entered what is now called Bass Strait and would have recognized Tasmania to be an island, not

214. Rossel. 1808. *Voyage d'Entrecasteaux*. 1: 209-210.

a peninsula. That discovery would await Lt. Matthew Flinders and his surgeon, George Bass, in 1798.

The necessary caution that dictated a direct route to the Tasmanian sanctuary of the previous year would ultimately fortify the suspicion of anti-colonialists that the French ministry of the marine had always had ulterior motives in dispatching the expedition to that region, thus accounting for the conscientious surveying of the Entrecasteaux channel. John Dunmore's observation, made over 30 years ago, that both French visits had been expedient, rather than planned, was based on evidence.[215]

The expedition reached the southern coast of Tasmania on 20 January 1793. Entrecasteaux had meant to proceed to the channel they had discovered the previous year, but Captain Huon asked that they enter the closest possible port. That had to be South Port in Recherche Bay, and there they anchored on 21 January 1793. The previous April they had seen snow on the coastal mountains; in January they were in the warmest season of the year, favorable for plant collections. A party from the *Espérance* went ashore to seek water, and a good source of easily-obtainable supplies was discovered in a small cove just to the northwest of the anchorage.[216]

215. Dunmore. 1965. *French Explorers in the Pacific.* 1: 311-312.
216. Rossel. 1808. *Voyage d'Entrecasteaux.* 1: 211-227; Labillardière. 1800. *Relation.* 1: 416-431.

XI
The Voyage d'Entrecasteaux: Part 5
Tasmania Revisited

Once the reprovisioning and the repair of both frigates was underway, Entrecasteaux sent a party to inspect the area of their of their anchorage the previous year in North Port. Their former watering-station was found to be completely dry, and Delahaye's garden had not suceeded. The General then asked Delahaye to visit the site to determine the cause of the failure. Had the garden been planted in the wrong season? or had the seeds not preserved their fertility? Labillardière now had the exceptional opportunity to collect again plants already known to him but in a different season of their development, as well as to find species overlooked in 1792. When he went ashore on 22 January 1793, however, his most memorable initial find was a skillfully-built aboriginal hut, its covering of tree-bark so arranged that even the hardest rains could not penetrate. From all appearances, it was a shelter where natives came to sleep at night. The desire to observe the interior of the hut led several in the party to enter it. The vermin inside they picked up on their clothes soon began to sting them, an unpleasant first lesson in the facts of aboriginal life.

At daybreak the next day, 23 January 1793, his party entered a substantial valley stretching to the southwest between hills of some height. From the top of one of them, they had a view of the entire country up to the foot of the high mountain, now Mt. La Pérouse, with its persistent snow, visible to the northwest of their ships. He took notes on the several species of *Embothrium* prominent on the slopes.[217]

Those slopes gave birth to periodic springs that drained down to the large plain below, turning it into a marshy, unstable terrain covered with the roots of various plants. Occasional small hollows would be covered with water, and around such places Labillardière made a substantial collection of plants. Notable among them were species of *Drosera*, an insectivorous genus.[218] On crossing grassier terrain, he sensed its suitability to the cultivation of most European plants, and the summer temperature was evidently favorable. Even the marshy ground could be drained with the proper ditching, rendering it very fertile land.

217. *Embothrium tinctorium* Labill. 1804. *Nov. Holl. Pl.* 1: 31-32. Tab. 42-43. [=*Lomatia tinctoria* (Labill.) R. Br.]; *Embothrium truncatum* Labill. Ibid. 1: 32. Tab. 44. [=*Telopea truncata* (Labill.) R. Br.]

218. *Drosera spathulata* Labill. Ibid. 1: 79. Tab. 106, fig. 1; *Drosera peltata* Labill. Ibid. 1: 79-80. Tab. 106, fig. 2 [=*Drosera peltata* Sm. ex Willd.]

He spent the entirety of the following day describing and preparing everything collected since their arrival in South Port, astonished by the number of species he had not found during a month of research the previous year. Accordingly, on the 25th, his party went out early, prepared to remain in the field for two days in order to collect at some distance beyond where Labillardière had just been. Going to the southwest, they picked up a trail used by natives, fresh prints made by bare feet quite visible. Soon they found the site of a recent fire.

In the afternoon, as they were approaching the southern cape of the island, a gusty wind brought in rain. A tufted shrub in the Rutaceae provided them an excellent shelter and proved to be, in Labillardière's opinion, a new genus he named *Mazeutoxeron*, the type-species being *rufum*.[219]

Turning back toward the frigates, northward near the sea and through sandy dunes, Labillardière found a new species of plantain that he called *Plantago tricuspidata* because of its leaf form, a species he would never publish. But at that moment it was prized as a useful plant, a salad green; and someone in the group had had the foresight to take along the necessary seasoning for such a food. Everyone enjoyed the tender leaves, and a plant had been found that could provide human nourishment. Among the grasses that seemed to stabilize the dunes, three appeared to him to be new.[220]

Despite two sleepless nights on that outing because of wind and rain, he devoted the next four days, 27-30 January 1793, to collecting in areas near the anchorage. To the south-southeast, he came upon a beautiful tree that appeared to be a conifer judging from the arrangement of its stamens and the resinous odor of all its parts. He was unable to obtain any fruit as the season was insufficiently advanced. The large, deeply-divided leaves were unique for a conifer in his experience. As an extract from them, bitter but aromatic, reminded him of spruce, he thought that, by adding malt to it, he could make an acceptable beer. His experiment worked, at least to his satisfaction. He added that the wood was very hard and could be worked to a high polish. Sometime later he would conclude that he had found a new species.[221] The deeply-divided leaves are now recognized as modified branches.

The thick forests to the north-northwest of their anchorage featured many trees of medium height and growing well despite the shade cast over them by the enormous trunks of *Eucalyptus globulus* Labill. Labillardière offered a descrip-

219. *Mazeutoxeron rufum* Labill. 1800. *Rel. Voy. Rech. La Pérouse.* 2: 11-12. Tab. 17. [=*Correa alba* Andr.]

220. *Festuca littoralis* Labill. 1804. *Nov. Holl. Pl.* 1: 22. Tab. 27; *Festuca pectinata* Labill. Ibid. 1: 21-22. Tab. 25 [*Agropyron pectinatum* (Labill.) Beauv.]; *Festuca scabra* Labill. Ibid. 1: 22. Tab 26 [=*Agropyron scabrum* (Labill.) Beauv.]

221. *Podocarpus asplenifolia* Labill. 1806. Ibid. 2: 71-72. Tab. 221 [=*Phyllocladus rhomboidalis* Rich.]

tion of one of them that reached not above 8-10m as a possible new genus in the Guttiferae, the Garcinia Family: its leaves oval, opposite, leathery, shiny, covered with a light coating of resin oozing from the upper side, and whitish below, the transversal nerves barely visible.[222]

On 31 January 1793, the naturalists developed a plan to visit the highest mountains in that part of Tasmania in the expectation of finding many new species. Each man carried food for five days, hardtack, cheese, salt pork, and eau-de-vie, the usual provisioning for distant excursions. Leaving early that morning, they followed their earlier route toward the southern cape until they reached the extended glade that facilitated their walk westward toward the hills. Once they entered the forested region, their only guide was the magnetic needle, and they bore back toward the north, immensely impeded by fallen timber. They reached the summit of a mountain after an extraordinarily difficult climb, from which they could see the tall mountain to the northwest that was their goal.

At nightfall, they built a large fire. They slept in the open, not taking the time to build a shelter, and not having seen any sign of natives since leaving the coastal region. "The air was entirely calm," Labillardière noted. "I awakened about midnight. Sensing myself isolated in the midst of those silent forests, their majesty revealed by the faint light of the stars, I was overwhelmed by a feeling of admiration for the grandeur of Nature quite beyond my power to express it."[223]

A second day of laborious walking, 1 February 1793, again brought them to a summit from which to view the tall mountain. For the first time, they recognized how much more distant it was than their initial calculation. It would evidently take their five planned days to reach the base of the mountain, at which point their food supplies would be exhausted. Turning back to the south-southeast, they encountered a substantial stream at the edge of the forest and elected to spend the night nearby.

Next morning they decided to follow the stream toward the sea, taking advantage of its neighboring glades and avoiding the fallen timber met on the ascent. The tactic worked although they still had great trouble with an escarpment they had earlier bypassed. Once they reached the coast and turned northeastward toward the anchorage, Labillardière found several species of *Lobelia* emerging from crevices in the rock, one of which still stands as a good species.[224] Night came on before they reached their destination. On the following morning, 3 February 1793, they encountered two large rock pools that featured an aquatic species

222. *Carpodentos lucida* Labill. 1800. *Rel. Voy. Rech. La Pérouse.* 2: 16-17. Tab. 18. [=*Eucryphis billardieri* Spach.]. Later placed in Eucryphiaceae.

223. Ibid. 2: 19.

224. *Lobelia gibbosa* Labill. 1804. Nov. Holl. Pl. 1: 50-51. Tab. 71.

with delightful flowers on the surface of the water.[225] The pools did not give off a fetid odor, suggesting the water was rapidly renewed by filtered underground water.[226]

After reaching the ships by mid-afternoon, he spent the remainder of the day describing and preparing the plants collected the previous days. He also spent at least a portion of the next two days walking over the low terrain to the southeast of the anchorage, finding an excellent soil for plants nearly everywhere. As he had been looking for a promising site for an additional garden for seeds brought from Europe, he made such a recommendation after returning to the *Recherche*. To his dismay, he learned that some very sandy, dry soil near the bottom of the port had been spaded and seeded during his absence, a site far inferior to what he had found.

On the morning of 7 February 1793, Labillardière and Delahaye, with two crewmen, set out to revisit the area beyond North Port, the previous anchorage, a two-day outing. The first day was uneventful; but on the second day, as they approached the large lake where they had botanized earlier, they heard human voices ahead. Moving cautiously and remaining concealed, they soon perceived a considerable number of natives engaged in fishing at the lake's edge. After a brief retreat to prepare some cartridges and settle upon a defensive strategy should it be needed, they again advanced in the hope that Labillardière would be able to communicate with the inhabitants.

Hardly had they gone a few steps when they met the aborigines, the men and boys drawn up in a semicircle, the women, girls, and infants somewhat behind. The formation indicated that the aborigines had anticipated, even welcomed, the encounter, and Labillardière would subsequently surmise that the aborigines had observed the sleeping Frenchmen near their fire the night before and had done them no harm when they had been quite unguarded. In any case, no hostile sign was given, and Labillardière approached the eldest male, who accepted a piece of hardtack after Labillardière had eaten a bit of it himself. He extended his hand as a sign of friendship and had the pleasure of receiving the native's hand and a pleasant smile.

The natives joyfully accepted neckerchiefs that were offered to them. As the previous night had been cold, the French were still wearing considerable clothing. They took off all they could spare in favor of the islanders, surprised that the islanders were entirely nude given the climate. There were 42 natives in the group: 7 adult men, 8 women, the remainder their children. They had woolly hair,

225. *Utricularia dichotoma* Labill. Ibid. 1: 11-12. Tab. 8.

226. Labillardière. 1800. *Relation*. 2: 6-24.

and beards were allowed to grow. Their skin was not a deep black. The practice of extracting two of the upper incisors, reported by earlier travelers as a common custom, was not evident within that group, all of whom had fine teeth. Being nude, these aborigines were subject to injuries that were especially visible on the lower extremities.

Knowing that the beach was a considerable distance, where a boat would soon be waiting for the French party, Labillardière reluctantly gave signs for departure, sorry to leave such peaceful people. To his astonishment, the men and four of the younger males gave indication that they would accompany the French. One of the strongest men turned aside into the brush to retrieve two spears where they had been hidden, apparently so as not to frighten the visitors. They were now to have an armed escort.

The pace of that singular procession was quite slow; and it seemed that the natives were not accustomed to sustained walking, as about every half hour they would invite everyone to sit down for a few minutes of rest. They quite clearly understood that the botanical party had come from North Port, as on two occasions, at points where the trail forked, they indicated the correct path to reach that port most directly. As they neared North Port, Delahaye recognized they were also quite close to his original garden and was determined to visit the site.

Taking advantage of a moment when the aborigines were sitting down, and hoping they would remain seated with the two crewmen for fear they might damage any vegetables that had succeeded, Delahaye and Labillardière turned aside and came upon the garden. One native was insistent upon following them. He examined the remaining garden plants, pointing to them in a manner that suggested he did not recognize them as native plants.

An earlier reconnaissance of that area had brought back a report that the garden had failed. Delahaye and Labillardière were sorry to see that only a few cabbages remained, along with a few potatoes, some radishes, cress, chicory, and sorrel, all of them in very poor condition. Labillardière surmised that the plants could have fared better had the garden been planted closer to a stream they had just seen to the west of the site. Why, he wondered, had Delahaye not planted at least the cress along the stream bank? Surely an oversight! It was the second time within a few days that Labillardière thought fit to relate a doubt about Delahaye's horticultural acumen. Delahaye gave Entrecasteaux the explanation that the seeds had simply been planted too late in the season.[227]

When they reached the beach, none of the natives could be enticed to visit the ship. Consequently, other members of the staff were eager to go ashore the

227. Ibid. 2: 25-39; Rossel. 1808. *Voyage d'Entrecasteaux*. 1: 229-233.

following day, 9 February 1793, to have a glimpse of what seemed to be a unique people. The myth of the noble savage had been of literary origin in the 17th century:

> I am as free as nature first made man,
> Ere the base laws of servitude began,
> When wild in woods the noble savage ran.[228]

That image had survived in literary circles in the 18th century, especially as a useful weapon in social criticism. But explorers and naturalists by then had been publishing distressing reports to the contrary. Entrecasteaux had read the claims made by Nicolas-Thomas Marion-Dufresne in his notes for *Nouveau voyage à la mer du Sud*, published in 1783, wherein he characterized the Tasmanian aborigines he had encountered as malicious. And Entrecasteaux had to know that Marion, after his visit to Tasmania in 1772, had been murdered by natives on New Zealand and cannabilized along with fifteen other members of his party. He knew that Captain Cook had been prevented from landing on Tasmania by armed, black-skinned aborigines, and that Cook had later met his death in the Hawaiian Islands. Certainly La Pérouse had commented on the falsity of the literary tradition. Could it be that Labillardière, in southeastern Tasmania, had been welcomed and protected by an aboriginal group of such sweet disposition as to verify the most sanguine of philosophical suppositions?

Several boatloads of curious visiters drifted past North Port on the 9th until smoke from a fire signaled the presence of aborigines. A few of the islanders came down toward the beach uttering cries of pleasure over the reunion. The fiddler, who had brought along his violin, hoped to arouse the same enthusiasm as his tunes had done among the natives of Buka Island the previous year; but his pride was quite wounded by the utter indifference exhibited by these natives.

Climbing to higher ground, the French soon came upon the main part of the native band. Each family was grouped around its own fire, making a meal of mussels as fast as they could be cooked over the coals. A species of bladder wrack, *Fucus palmatus* L., was being grilled on some of the fires. When it had acquired a certain degree of softening, it was torn into pieces for eating. The natives, even if quite willing to let the French taste the natives' food, absolutely refused to taste any of the French food offered to them, including the French beverages. Their diet was evidently limited to what they had known from childhood.

The separation of families during eating allowed the observation that a man might have two wives, each one surrounded by her own children. As there were

228. John Dryden. 1669-1670. *The Conquest of Granada*. Part 1. I, 1.

no men without wives, the practice seemed to ensure that no woman would be without a husband and children. It also might be, he observed, that only the dominant males benefited from that apparent shortage of adult males.

Labillardière quite sensed the importance of observing even the slightest actions of a people who seemed to him to be so close to the state of Nature. He made note of the punishment administered by a father to one of his children for having thrown a stone at the back of an even younger child. He struck him quite gently on the shoulder, provoking a few tears, but preventing a recurrence. Occasional spats among children were smoothed over without violence but with a gentle correction, to Entrecasteaux's great admiration. But in describing that primitive band "as having no other laws than the laws of Nature,"[229] he saw less clearly than Labillardière, who understood the implication of punishments for transgressions. Primitive they might be, living close to Nature without doubt, but they had a social order. Some aspects of it, indeed, less palatable to the taste of enlightened gentlemen, had yet to be revealed to them.

When it came time to return to the boats, more than half of the peaceful inhabitants got up to accompany the French party to the beach. On this occasion, four girls joined the group, three of whom were described as already nubile. The clothing distributed to the natives the day before had been casually shed as a useless burden. The men followed along with a measured gait, hands often folded behind their backs. The young girls were of a more playful temper. One of them talked continually, speaking for a very long time with an extraordinary volubility. The French found her performance curious on the assumption that she had to perceive that they could not understand her, but that she simply had to talk. That assumption was probably erroneous. How many of us have had experiences with unsophisticated foreigners who believe they can make themselves understood merely by raising their voices and talking faster!

The visitors were immensely conscious of how much they were missing because of their inability to understand the local language. Most of the critical information about that group of aborigines could not be obtained: Why had they eschewed any association with the visitors for weeks, only to come forward belatedly in such an amiable manner? Was their geographical isolation from other peoples as great as their peacefulness implied? That is, was armed conflict unknown to them? Or such questions as what did they know of their past? Or did they have any concept of time and the past? Beyond which, what questions did the aborigines have about their visitors? What impressions were they making upon the inhabitants?

229. Rossel. 1808. *Voyage d'Entrecasteaux.* 1: 234-235.

By 10 February 1793, the French were in the final stages of repairs to the frigates and in charting the large bay beyond their port. All installations on shore were abandoned that date. A considerable number of personnel from both ships went ashore in an attempt to see the aborigines once again. A few natives came down to the boats, then guided them along shore for at least 2k until they found themselves in the midst of 48 aborigines. Their cooking fires had been lighted, one for each family.

Two of Piron's surviving illustrations, although undated, were obviously inspired by the events of that day. The first drawing was entitled *Sauvages du Cap de Diemen Préparant Leur Repas*, which, seventy years later, would surely have been called *Déjeuner sur l'herbe.* The second drawing, *Peche des Sauvages du Cap de Diemen*, suggests a charming frolic at the water's edge. Beyond the misleading neoclassical portraiture in the drawings, they quite failed to depict less uplifting occurrences of that day.[230]

The violinist, despite his earlier failure to enthrall this primitive audience, had since developed the conviction that the aborigines would respond to more lively tunes played with a pronounced rhythmical beat. He had brought his fiddle for the attempt. At first, they seemed to listen in some uncertainty, leading the fiddler to redouble his efforts to win their applause. But he soon saw the numerous assembly begin to put fingers into ears to avoid hearing any more, and the bow fell from his hands.

It was also noticed that the natives were infested with vermin. "We admired the patience of a woman," Labillardière wrote, "who devoted considerable time to freeing her children of them. But we also saw with considerable distaste that she crushed those disgusting insects with her teeth, swallowing them immediately. We could not avoid remarking that monkeys have the same habit."[231]

Inasmuch as the sex of an individual, in a country where complete nudity is the rule, cannot ever be in doubt, the local aborigines had been confused by the appearance of fully-clothed Europeans. They could not detect which ones among them were female. Just how that confusion was conveyed to Labillardière we are not told; but he had come to realize that, as the natives could not accept the explanation that all the Europeans were men, they had concluded that the youngest members of the staff and crew were actually women. Their curiosity in the matter went far beyond anything Labillardière had expected, and they could not be convinced of the truth until obtaining assurance of that fact by a convincing exposure.

230. Labillardière. 1800. *Relation*. Atlas, Tabs. 3-4.
231. Ibid. 2: 50.

But the visitors had only partly appreciated the essential role of women in that aboriginal society. Until then, they had had little idea about the efforts made by the women to provide food for their families. Toward midday, they watched the preparations for a meal. In the words of Labillardière: "Each woman took a basket and, followed by her daughters, went out on the rock that projected into the sea. From there, they ventured to the bottom of the water in search of crustaceans and shellfish. As they remained under water for a long time, we became quite anxious for their fate, aware that they had dived into marine plants of considerable length. Finally they reappeared, having proved to us that they could remain under water more than twice as long as our most skillful divers.

"They required only a brief moment to breathe before diving again until their baskets were nearly full. Most of them used a small spatulate piece of wood to detach mollusks from rocks hidden under water and the very large abalone at greater depths, apparently their favorite. At the sight of large lobsters in their baskets, we feared such crustaceans would rip the unfortunate women with their claws, but soon realized that they had taken the precaution of killing them as soon as they had been captured. The women emerged from the water only to bring the fruit of their fishing to their husbands, often returning to dive again until they had procured sufficient abundance to feed their families.

"Each one of us was profoundly stricken to see those poor women condemned to such arduous work and exposed to such dangers as sharks and entangling seaweed, while the men sat around their fires, enjoying the best of the cooked food, also eating grilled rockweed and fern roots. Several times we urged the husbands to at least share in the laborious work, but in vain. Occasionally they would break a branch into small pieces of dry fuel to feed a fire, but that was the limit."[232]

The meal went on for a considerable time without anyone having anything to drink. Once everyone was entirely sated, the women and girls went to find water with a type of seaweed pot they had occasionally seen earlier. Water was drawn from the nearest source and brought to the men. They drank it without repugnance no matter that it was stagnant and very murky. That terminated the meal.

The received idea that freedom and goodness would be found close to the state of Nature was, for those who witnessed that meal, somewhat less certain after lunch than it had been before lunch. While they had seen no sign of any chiefs, each family apparently living independently, it had been noted that the children were definitely subordinate to their mothers, and wives entirely subordinate to

232. Ibid. 2: 52-54.

their husbands. Primitive the society may have been, but it was unquestionably authoritarian with no rights of appeal.

The plight of the women, a French perception to be sure, generated further speculation about the effect of polygamy on the women. Several of the more robust men had two wives, the remaining men only one. Which status, from the female point of view, was more enviable? In the cases of polygamy, the women shared the most burdensome domestic work, but they also shared in demonstrations of their husband's affection. Although the issue was not resolved by the French observers, it would not have arisen without the recent revelation of the subservience of women. Labillardière fell short of using the term *slavery*, he would retain a favorable memory of those amiable aborigines, but any notion of a paradise had become insupportable.

Inasmuch as Admiral d'Entrecasteaux would insert a glowing portrait of those Tasmanian natives in his journal, it needs to be remarked that he had not been witness to the natives' repast on 10 February 1793. In fact, he had not seen any natives at all before a final encounter with them on 12 February 1793. On that morning, five natives appeared on shore and built a fire to attract attention. Several boatloads of the French responded with presents for the delighted islanders. One was given an axe and shown how to strike in order to make the cut that would fell a tree. He was nearly overcome with joy when the tree came down. The natives were astonished by the rapidity with which its trunk could be sawed through, and they were given several handsaws and taught to use them. As it had become apparent they did not have fishhooks, they were given some and shown how they were to be used, the French congratulating themselves for having provided the means of alleviating one of the most difficult tasks assigned to women.

The aborigine who had learned to use the axe quickly was induced to examine the two frigates through a good telescope. He soon yielded to invitations to get into a longboat to visit the *Recherche*. He seemed quite assured on board, examining the interior of the ship with great care. The livestock on board seemed to astound him. He had seen nothing analagous to it except for the guinea fowl, which had a similarity to the black storks of Tasmania. He asked for one of them, indicating an intention to eat it. He was especially struck by a handsome rooster in one of the chicken cages. That, too, was given to him for the same purpose. After about 30 minutes, he asked to be returned to shore and was at once accommodated.

Several of the naturalists had taken pains to record words from the aboriginal vocabulary, starting from the initial encounter. The final interview provided the greatest opportunity to add to that vocabulary, by which time the botanists had been able to discern that the natives had a special name for each plant. Different individuals were consistent in providing the same name for a given plant. Although Labillardière did not say so, the chances are that such distinctions were made at the generic, rather than the specific, level. Brief vocabularies would be published in both the principal narratives of the voyage.[233]

Entrecasteaux's reflections upon the character of the Tasmanian natives portray reliably the affection the French had come to feel for those natives. Having seen less of them than Labillardière had, Entrecasteaux expressed a judgment at once less realistic, but more compatible with those visions of noble savagery dear to utopians in Europe who had not seen them at all:

> "The group we have seen seemed to provide the most perfect picture of the initial stage of society when men are not yet disturbed by passions or corrupted by the vices sometimes induced by civilization. Composed of a few families gathered together, without any other property than their wives and children, they are devoid of any cause for dissention. Having no chiefs other than those Nature itself designates, that is to say, the fathers of the family and the elderly men, there has to exist a mutual affection among the men who compose the group; and whose ties are tightened even more by the marriages they are obliged to contract from within the small number of families that are flocked together.
>
> "What is more, these men, assured of finding their subsistence easily, must enjoy peace and contentment. Did not their physiognomy, open and smiling, suggest a picture of a happiness untroubled by obtrusive thoughts or by futile desires? Less advanced, no doubt, than the peoples of New Zealand in civilization, they also lacked their ferocious disposition. The hatreds that divide the inhabitants of that island appear to be unknown to those on this part of Australia."[234]

The two frigates had been prepared for departure for several days. At daybreak on 13 February 1793, they weighed anchor, meaning to move out of their bay and then northeastward into the Entrecasteaux Channel. The immediate goal was to round the north end of Bruny Island in order to sail southward along its east coast to reach Adventure Bay where Captain Cook had taken on water. Contrary winds and currents in the channel obstructed progress and forced them to drop anchor until conditions improved.[235]

233. Labillardière. 1800. *Relation*. Appendix 2: 38-40; Rossel. 1808. *Voyage d'Entrecasteaux*. 1: 552-556.
234. Rossel. 1808. *Voyage d'Entrecasteaux*. 1: 242-243.
235. Ibid. 1: 233-240, 248-259; Labillardière. 1800. *Relation*. 2: 55-64.

The halt gave the naturalists an opportunity to land on the west coast of Bruny Island. Among the shrubs, Labillardière sighted one that appeared to belong within *Mazeutoxeron*, a genus he had named earlier [=*Correa* Andr. in Rutaceae]. All the characters placed it there; but the petals, separate at base, seemed to adhere at mid-section, yet could be separated without breaking them. The style was simple and acute; flowers greenish, solitary, subtended by two small oval leaves; two filiform appendages midway on the peduncle. He gave the new species the name *reflexum*, because thc leaves, villous and whitish below, were bent downward.[236]

Adverse weather conditions prevented the ships from advancing for the next several days. On 16 February 1793, some natives appeared on the shore of Bruny Island and lighted a fire. From time to time, they responded to calls from the crewmen with shouts that suggested a friendly disposition. Finally, a considerable party went ashore. The natives advanced, unarmed, expressing evident pleasure over the visit.

Using words that the naturalists had learned from the aborigines on the mainland, the natives were invited to sit down. They understood and complied at once. It soon became clear that they also understood other words from the mainland vocabulary, leaving no doubt that both groups spoke the same language. The matter was perplexing, because William Anderson, with Cook, had earlier reported on words spoken by inhabitants around Adventure Bay, on the east side of the island, that had no relation to the vocabulary Labillardière was in a position to verify. He was already aware, however, that there seemed to be no connection between the vocabulary he had been assembling and several vocabularies reported by earlier voyagers from the east coast of Australia. The peoples of that region, he had concluded, were not all of the same origin.

One of the natives got the naturalists to understand that he had previously seen ships in what had to be Adventure Bay, mostly likely those of Captain William Bligh. A few days later, in fact, the French would find inscriptions carved into tree trunks at Advanture Bay proving that Bligh had anchored there early in 1792.

The aborigines of Bruny Island were greatly similar to those previously visited across the channel, except that nearly all the men were rather elaborately tatooed in straight lines across the chest. Neither group seemed to know the bow, having only spears that were concealed in the nearby woods; both groups lived primarily on shellfish. These natives showed great gratitude when presented pieces of variously-colored cloth, glass beads, some ironware, and an axe.

236. ***Mazeutoxeron reflexum*** Labill. 1800. ***Relation.*** 2: 66-67. Tab. 19 [=***Correa speciosa*** Ait.]

The French party returned to the ship just before sundown; and, as the wind had become favorable, the ships were able to proceed about 10k before dropping anchor for the night. By early morning, as the wind had again become contrary, it was possible to botanize again on Bruny Island by following native trails into the forest. Labillardière soon observed a new species of his *Exocarpus* that he called *Exocarpus expansa* Labill. Its branches were more distant, one from another, than those of *Exocarpus cuppressifolium* Labill., and its fruit was larger than that of the latter species.[237]

Two cannon shots from the *Recherche* announced imminent departure and cut the field trip short. The breeze proved to be so feeble that little headway was made, and it took four more days to clear the channel. They finally anchored in Adventure Bay late in the morning of 22 February 1793. During the few days they were at anchor, while the crewmen sought the source of water indicated on Captain Cook's map, Labillardière set out daily to seek plants.

The results were disappointing. He found little he had not already collected on the Tasmanian mainland. On Penguin Island, which proved to be really an extension of Fluted Cape (the arm protecting Adventure Bay on the south), he was pleased to find a new eucalyptus of medium height, easily recognized by distinctive leaves: opposite, sessile, and glaucous.[238] He also found the embothrium with very deeply-cut leaves,[239] as well as several species of what he thought were odorless *Philadelphus* that he neither described nor published.

Evidence of Captain Bligh's earlier visit to Adventure Bay was soon found on shore: the remains of a platform for sawing wood, pickets used for the support of a tent, and some large blocks that had probably served as bases for astronomical instruments. Several inscriptions carved into tree trunks indicated that Bligh had anchored there in February 1792. He would have been on his way to Tahiti to obtain breadfruit trees for transfer to British colonies in the West Indies.

A marker was found indicating a planting of seven fruit trees, the inscription reading that the planting had been made by "Wm. Bligh and S. and W." Delahaye was sent ashore to examine the state of the trees. He reported the existence of two pomegranate trees, one quince tree, and three fig trees. Shoots from the trees were about one inch long, but appeared to be vigorous. The trees themselves were quite small, perhaps because they had become crowded by ferns surrounding them. A little beyond that planting, an apple tree was found that was nearly 2m tall, the trunk 5cm thick, suggesting it had been planted several years before Bligh's visit.

237. Ibid. 2: 73.
238. *Eucalyptus cordata* Labill. 1806. *Nov. Holl. Pl.* 2: 13-14. Tab. 152.
239. *Embothrium tinctorium* Labill. 1804. Ibid. 1: 31-32. Tabs. 42-43 [=*Lomatia tinctoria* (Labill.) R. Br.].

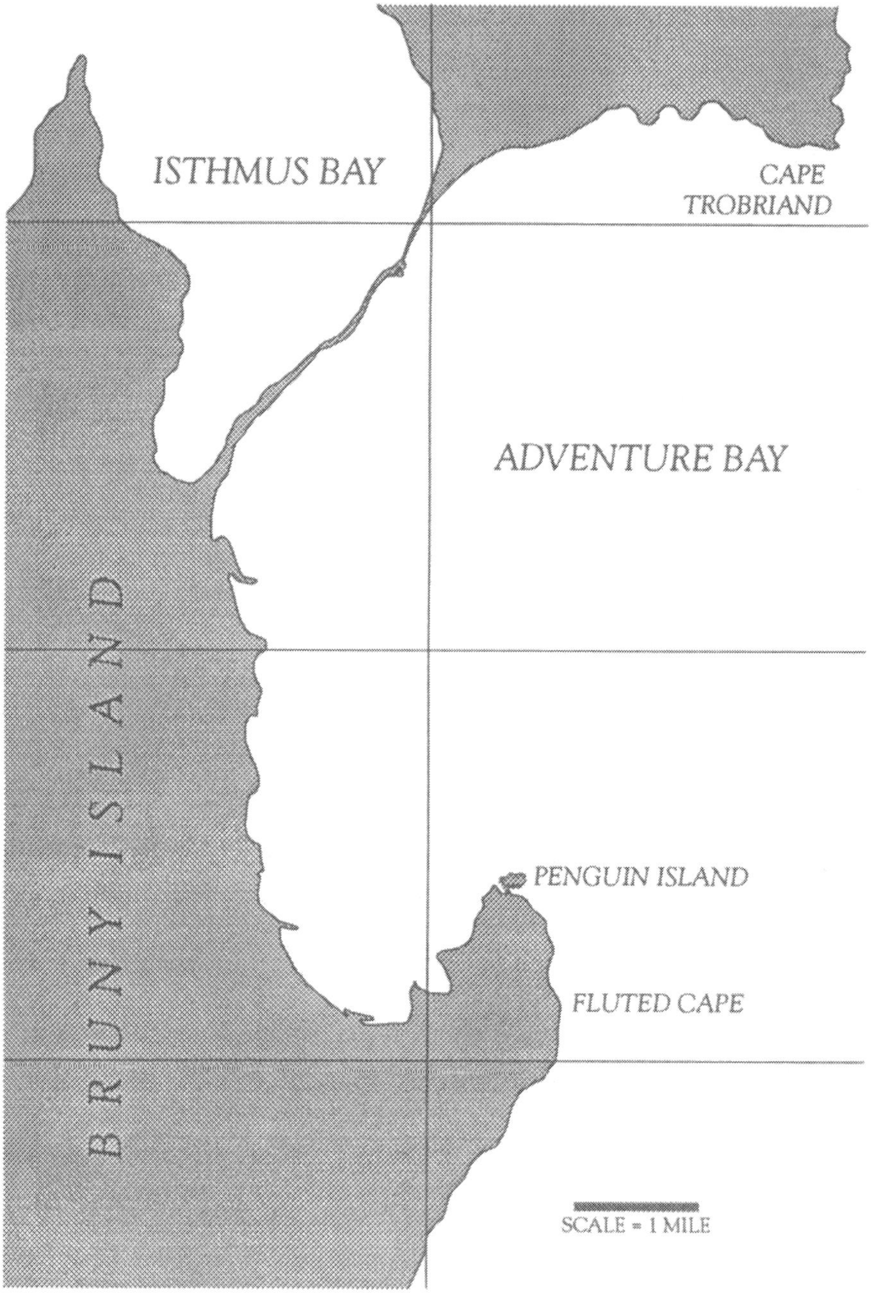
ISTHMUS BAY
CAPE
TROBRIAND
ADVENTURE BAY
BRUNY ISLAND
PENGUIN ISLAND
FLUTED CAPE
SCALE = 1 MILE

It did not appear to have been grafted, was not doing well, and a long life for it was doubtful.

A second marker indicated a planting of a garden cress, but there remained no trace of it. A third inscription marked a planting of cabbage, but there was no evidence it had succeeded. Nor was there any evidence that the sow and boar left by Captain Cook had survived and multiplied. Even so, a pregnant goat and a young buck were put ashore on the northern side of the bay in the hope that the aborigines would allow them to propagate on Bruny Island.[240]

Perhaps, Labillardière speculated, the goats "could multipy to a point of provoking a total change in the way the inhabitants live. If able to become a pastoral people, they would abandon the seashore without any regret and would enjoy the pleasure of no longer being obliged to dive for their subsistence at the risk of being devoured by sharks. The women, being condemned to such difficult work, would sense the value of such a present even more than the men."[241] He feared, to be sure, that these quadrupeds would be killed before reproducing, even though there was no sign of recent native habitation on that coast of the island.

Labillardière's observation, based upon differences in vocabularies, that not all of the people along the eastern coast of Australia were of the same origin, would be verified by scholars in the 19th century. Although the Tasmanian aborigines visited by the Entrecasteaux expedition would be extinct by 1876, sufficient observations of them had been made to establish that they had evidently been of Papuan origin and that their language belonged to the Polynesian group. The blacker Australian aborigines represented a race distinct from the Tasmanians, possibly a Dravidian people driven at some time in the distant past from their primitive home in the Indian Deccan. They spoke a variety of dialects of the same language, which had no affinity to the Polynesian group.

The Tasmanian population cannot have been large in the 18th century. Estimated to have been 5,000 at the very most, it did not long survive in the contest with European settlers that began in 1802. By 1842, only 44 of the aborigines could be counted; and the last pure-blood Tasmanian died in 1876, a woman said to have reached about the age of 76. It was appropriate that Bruny Island became the final refuge for those Tasmanians before their complete extinction, as it had been named for the man who would have become their protector had he lived.[242]

240. Rossel. 1808. *Voyage d'Entrecasteaux.* 1: 259-267.
241. Labillardière. 1800. *Relation.* 2: 79.
242. Dunmore. 1965. *French Explorers in the Pacific.* 1: 313.

Portrait of Jacques-Julien Houtou de La Billardière.
© 2000, Hunt Institute, CMU, Pittsburgh, PA 15213, U.S.A.

XII
The Voyage of Entrecasteaux: Part 6
On to New Caledonia

A sudden change in the wind on 27 February 1793 enabled the frigates to get underway. Entrecasteaux meant to resume his attempt to follow La Pérouse's projected route. That required the Tonga Islands to be the next visit. Ample provisions would be readily available on islands so heavily cultivated, and the relative advancement of the native population offered some assurance that, if La Pérouse had indeed visited the islands, reliable information could be obtained. Conversely, the Tonga Islands could no longer be a happy hunting ground for botanists, who would have to occupy their days observing native mores. Frank Horner's recent description of the Tongan interlude, colorful and unequivocal, need not be duplicated here.[243]

Upon leaving the Tasmanian coast, the two frigates headed for the northern tip of New Zealand. The Three Kings Islands were sighted at daybreak on 11 March 1793 and New Zealand later that morning. When they approached North Cape later that afternoon, several native pirogues came out intent upon trade. The men were very tall and strongly-muscled, demanding axes in particular in exchange for their wares and fresh fish. The French obtained some of the bones the natives wore over their chests, suspended from a cord around the neck; the bones were soon identified as small parts of the human forearm, the ulna.

As the ships were still close to shore the following morning, Labillardière believed it to be his duty to remind Entrecasteaux how important it would be to take back to France the liliaceous plant from New Zealand, *Phormium tenax* J. R. & G. Forst., the New Zealand flax. He was sure it would thrive unquestionably in Europe and explained that the fibers removed from its leaves have a strength quite superior to those taken from any other vegetal products used to make cord or rope. Cables made of it would resist the greatest stress, a matter surely to interest a naval commander of Entrecasteaux's quality. But the sad fate of Captain Marion-Dufresne on New Zealand in 1772 (ambushed, murdered, and devoured along with 16 crewmen) made the General determined to move on.[244]

After reaching the Tonga archipelago, the French anchored in Tongatapu harbor on 23 March 1793, where they would remain until 9 April. Although the island group had been called the Friendly Islands by Captain Cook, Entrecasteaux

243. Horner. 1995. *Looking for La Pérouse*, pp. 146-155.
244. Labillardière. 1800. *Relation*. 2: 82-88.

was quite aware that earlier visiters had experienced violence and meant to remain prudent and wary. The people in the swarm of pirogues that came out loaded with edibles seemed friendly enough, but the General thought it would be safer to set up camp, not on the main island, but upon a small island nearby from where to conduct scientific observations and trade.

He had an additional anxiety. As Labillardière put it, "We noticed that one pirogue contained no produce and had nothing to offer us. We were mistaken. This pirogue was directed by two men whose beaming faces announced their pleasure in pointing out to us two women who were paddling with them. The signs they made left no doubt that the merchandise being offered was love."[245] The General, very concerned to prevent the spread of disease, issued orders prohibiting that commerce. For whatever reasons, the sentinels posted could not cope with the many girls who easily escaped their surveillance by continually slipping through gunports or scuttles. We are given the impression that prostitution had become, at the very least, a minor industry on Tongatapu and perhaps accounts for Cook's designation, the Friendly Islands.

Trading began on the small island on 25 March 1793 under the close supervision of Lt. Dominique de La Grandière from the *Espérance*. It began quietly enough, but attracted an increasing number of noisy, armed natives as the day wore on. The danger of an attack became so obvious that Entrecasteaux ordered a cessation of trade and, finally, an evacuation of the camp, carried out under protection of armed guards. Trading did not resume until the 28th, only after the various chiefs had been convinced that all trading depended upon them maintaining order. Even so, subsequent trading between pirogues and frigates remained worrisome; partly because natives persisted in clambering aboard ship in order to steal anything portable, partly because it facilitated prostitution. Ultimately there was violence with several of the French being seriously injured, and two, or perhaps three, of the natives killed.

Why had the French remained in the vicinity so many days when the hostility of armed males had become so evident? Entrecasteaux knew that an explanation would be expected from him. He had anchored at Tongatapu primarily to discover whether La Pérouse had anchored among the Tonga Islands, only to recognize that the English equivalents of the Tongan vocabulary, assembled by the Cook expedition, would not serve the French. The naturalists, therefore, needed time to learn enough Tongan vocabulary to make themselves understood. After gathering information from various islanders whom they considered to be the most

245. Ibid. 2: 93.

intelligent, it appeared beyond any question that La Pérouse had *not* visited any of the major Tonga Islands.

All of Cook's visits had been remembered, along with the intervals between them, as well as was Captain Bligh's subsequent stop at Anamuka Island. Had any large ship called at Vava'u Island to the north of Anomuka, the news would have been brought to Tongatapu. The medallions that Entrecasteaux's people had been distributing to some of the natives were identical to what La Pérouse would have distributed had he called, but they seemed to be entirely new to the Tongan recipients.[246]

The General, meanwhile, had been endeavoring to discern the structure of power, or the character of authority, on the Tonga Islands. He made the arresting observation that the widespread *abuse* of authority, that is, the *right* to demand from a social inferior anything that inferior might possess, could be the key to understanding the disposition to theft so characteristic of the peoples of the South Seas: the *right* to take anything you have the power to take or the opportunity to take. The natives had demonstrated a dexterity equal to the pickpockets of the European capitals, an accomplishment suggesting long habit. Foreigners, no doubt, were more vulnerable to thievery as they possessed items of great value to such islanders. It had been evident everywhere that pilfered items often found their way to the chiefs, an explanation for why chiefs were solicitous to see that trade was not interrupted; and why their efforts to forbid thievery were largely theater meant to dampen European outrage.

Entrecasteaux by no means found the Tongans ferocious in character, if anything, naturally disposed to merriment and mockery. But most fundamental in their character was the apparent absence of any conception of humane feeling: no feeling for, or sentiment about, humanity. He found himself endeavoring to depict a society without justice, in which thievery had been ingrained in the character of the people, rendering them dissimulating and treacherous.

He was pleased to add that at least the women were better treated on the Tonga Islands than elsewhere in the western Pacific where the most difficult work was reserved for women. Here the women were meant only for household duties and the instruction of children. Consequently, they seemed to enjoy better health and to have a more vigorous constitution. Both men and women, at least of the upper social ranks, maintained an extreme cleanliness.[247]

Entrecasteaux gave high priority to collecting breadfruit plants on Tongatapu and getting them back to Mauritius in a good state. From Labillardière's descrip-

246. Rossel. 1808. *Voyage d'Entrecasteaux*. 1: 268-302.
247. Ibid. 1: 303-310.

tion, it seems evident they acquired suckers from cultivated breadfruit trees, *Artocarpus altilis* (Parkinson) Fosberg, rather than wild stock. The fruits, about 3 x 2dm in size, were seedless, whereas the fruits from wild species were much smaller, fewer in number, and contained large kernels that were difficult to digest. The local practice was to propagate the plants from shoots. A breadfruit tree may produce fruit for eight months of the year, an abundant and healthful nourishment for islanders, accounting for the zeal to naturalize it in the West Indies.

About 300 cuttings were taken aboard each ship. The general provided Labillardière and Delahaye space over the rest rooms on the port side, as well as a large chest in which to pack their stock. A good compost was obtained from the same natives who had provided the shoots. As an experiment to determine the best method of shipping the cuttings successfully, some were planted upright in the compost, the remainder buried horizontally in a loamy soil. Labillardière and Delahaye expected to plant the cuttings once they reached Mauritius.[248]

As Entrecasteaux had become convinced that La Pérouse had not called at the Tonga Islands, once the breadfruit cuttings had been stored, the final loading of fruit, vegetables, pigs, and fowl was undertaken in preparation for sailing. One final incident remained to bemuse the General, as it seemed so out of character with conventional Tongan behavior. The French had frequently seen chiefs openly seize the property of commoners and had always been struck that this sort of oppression seemed in no way to lessen the happiness of their character. But Koué, a handsome man and the eldest son of Queen Tinée, had none of the lighthearted-ness characteristic of the islanders.

Several days before the departure of the frigates, two of the Tongans, both of them men of high social rank, began to show an interest in joining the ships and leaving the islands. But Koué alone proved his serious intent by refusing gifts, wanting nothing more than a space on board; and Koué alone went on board the *Espérance* the day before departure intent upon remaining. Captain Huon had him transferred to the *Recherche* to leave the decision to Entrecasteaux. The latter tried to dissuade him, arguing that he would be exposed to the fatigue and weariness of a difficult navigation, that he risked the probability of never seeing his homeland again. But nothing changed his mind, either because he could not understand the General's arguments or could not really comprehend the full implications of his decision.

With considerable misgivings, Entrecasteaux then gave his consent, and Koué returned to the *Espérance* seemingly satisfied. During the noon meal, when he was seated with the command staff, several pirogues arrived bearing Tongans who

248. Labillardière. 1800. *Relation*. 2: 163, 168-169.

were close to him, sent by the queen to beg him to come ashore to bid goodbye, his last goodbye, to his family. (He had nine wives and nine children.) He could not refuse, saying that he would return in the afternoon. He was seen no more, having apparently met firm opposition to his departure.

Entrecasteaux was not sorry: "The man, of very noble countenance, but very serious-minded and melancholic, could only have resolved to leave his homeland for sorrows that the sojourn on the ship would certainly not have dissipated. How could he have supported the memory of having left nine wives and as many children behind? His regrets and pain would increase with every moment, and there could be no remedy."[249] We know not whether Koué changed his mind or was detained under guard. But more to be regretted was his inability to explain what he had seen that had made him certain that a better life beckoned over the horizon from his tropical paradise.

The expedition sailed from Tongatapu early on the morning of 9 April 1793 bound for the northeastern side of New Caledonia where they had been denied a visit the year before. On the 15th, they sighted the Ile Erromango in the New Hebrides; and two days later, when passing through the archipelago now known as the Loyalty Islands, they named two wooded islets amidst reefs, the Beautemps-Beaupré Islands. Later that same day, they could see a high mountain on New Caledonia, no doubt Mont Panié, its summit shrouded in clouds.

At daybreak on 18 april 1793, they approached to within 1.5k of the barrier reefs, then sailed northwest parallel to them in search of Balade Pass, the passage between the main island and Balabia Island, where Cook had found a safe anchorage in 1774 by what he called Observatory Island. Because of difficult winds and currents, they were unable to reach that destination until 21 April. During the interim, the frigates were approached by groups of New Caledonian aborigines in pirogues. The French held up yams and coconuts to indicate what they would like to acquire, only to discover that the natives themselves wanted coconuts and yams, offering spears and clubs in return. They rubbed their very flat stomachs to indicate hunger. Most of the men carried a small bag on their waistbands filled with stones, cut oval-shaped, that they could hurl with a sling.

By the time the frigates had anchored about a kilometer off Observatory Island on the morning of 21 April 1793, Labillardière had discovered a distressing situation. He had watered his chest of breadfruit cuttings the day before, yet noticed water dripping again from the chest as if recently irrigated. He became convinced of a malevolent deed when he tasted the dripping water and found it

249. Rossel. 1808. *Voyage d'Entrecasteaux.* 1: 314-315.

to be seawater. The cuttings could not survive the salt treatment, leaving only one chest on the Espérance. His failed to discover the identity of the offender.[250]

There were additional unsettling revelations that day. The French officers were familiar with Johann Georg Adam Forster's account of Cook's visit in *A Voyage Around the World* (1777), where the New Caledonian natives had been portrayed as gentle and benevolent, indicating that no serious protective measures were necessary. When the French went ashore that afternoon, they were soon surrounded by natives wanting to exchange their personal effects for cloth and iron. The natives quickly proved to be highly insolent rascals, being bold, experienced, adroit thieves like those on Tongatapu. The trading was carried on in bad faith, a native often fleeing before surrendering a promised item.

This initial bad impression was reinforced when an aborigine, a recently-grilled bone in his hand and eating a morsel of meat still attached to it, walked up to Piron, the artist, and offered to share the meal with him. Piron, presuming he was being offered a piece of some quadruped, accepted the bone, but handed it to Labillardière for identification. Labillardière found it to be the pelvis of a child, perhaps 14 or 15 years old. Other aborigines around him conveyed the impression that the bone had provided a very delicate food. The gnawed bone was taken on board the *Recherche* where the chief surgeon confirmed it as having belonged to a child. Other similar artifacts were soon observed, but the natives endeavored to give assurance that they ate only enemies.

Entrecasteaux gave orders that wood-cutting and water parties would henceforth be heavily guarded.[251] Upon going ashore the following morning, 22 April 1793, Labillardière's party immediately encountered aborigines already taking their meal. The recently-grilled meat was undeniably human flesh. With many expressive gestures, the natives not only indicated that the slice of meat had been cut from an arm, but that the man had been killed in a fight.

Continuing inland in search of additional sources of good water, they saw some plantations of yams and sweet potatoes. The flora, however, seemed disappointingly reminiscent of plants already collected in Tasmania. Moving into higher elevations, they were surprised to come across low walls built in upward tiers, a terracing designed to control the slippage of cultivated terrain. Labillardière had found a similar practice among mountain dwellers in Asia Minor. At some point, the few natives who had been following along refused to go farther, indicating dangerous country ahead where the party would be vulnerable to attack.

250. Labillardière. 1800. *Relation*. 2: 181-188.

251. Rossel. 1808. *Voyage d'Entrecasteaux*. 1: 331-334. Readers familiar with recent controversies about the validity of earlier reports of cannibalism should ponder the intelligent appraisal of the dispute by Paul Henley, "A Generous Quantity of Fat." *London Review of Books*. 21, no. 17 (2 September 1999): 24-26.

In fact, upon returning, they learned that there had been threats of violence that day along the beach, proving that the danger was everywhere, not just beyond the mountain.

Accordingly, the naturalists went ashore on the 23rd with armed crewmen assigned to cut wood and did not stray far from them. But Guyard, the master cannoneer on the *Espérance*, ventured into the forest to hunt for game. Toward noon, he perceived over 200 aborigines practicing spear-throwing while engaged in various maneuvers. He withdrew unseen and brought the news to the wood-cutting party.

Their axes had been stacked while they were eating their noon meal. The armed natives soon appeared, obviously intent upon seizing the axes. An officer ordered the removal of the axes to the longboat, but the natives assailed the boat in their determination to obtain axes. Several shots were fired, and one of the boldest of the aborigines was shot down, yet had the strength to drag himself into the forest. The remainder retreated into the cover of the forest from where they launched a hail of stones with their slings, but injuring no one. The action caused Entrecasteaux to have two cannon shots aimed between the natives and the water source. The cannon balls raised great showers of sand, and the natives fled. Although the French thereafter would not relax their guard, it would appear that the demonstration had taught the natives the inadvisability of attack.[252]

Whether or not the aborigines of New Caledonia limited their cannibalism to enemies killed in war, Labillardière recognized that they all revealed a habitual shortage of food. Beyond their evident flat stomachs, they seemed to give minimal heed to the cultivation of their land which indeed appeared to be quite infertile. Most of the natives who approached the French would ask for food, and they did eat the hardtack offered. Labillardière saw one native eat the local soapstone, a greenish clay that was quite friable but provided no nutrients. When others were seen to eat copious amounts of it, he deduced it was eaten to moderate the sense of hunger. Would cannibals, he wondered, have recourse to such an expedient when pressed by hunger?

When consequently, on 24 April 1793, Labillardière and Delahaye ventured into the forest, they took an escort of only two crewmen. Delahaye spread seeds of various European plants as they walked through the forest. What remained, he gave to an elderly aborigine who seemed to understand that he should plant them. Labillardière believed he recognized *Commersonia echinata* Forst., a tree growing abundantly in the Moluccas. Among the species he thought to be new

252. Ibid. 1: 334-335; Labillardière. 1800. *Relation*. 2: 191-203.

that day was a jasmine with sulphur-yellow, odorless flowers, and simple leaves. He would not publish it later.

Naturalists from both frigates worked out a plan to explore the back sides of the mountains to the south of their anchorage on 25 April 1793. The party numbered 28, and it was agreed that each man would carry arms in case aborigines dared to attack. They followed well-worn trails for a considerable distance into the mountains where the inhabitants appeared to live in the greatest distress, all of them extremely thin. As they gained in elevation, they noted that the nearly constant east-southeast wind inhibited plant growth. Trees common to lower elevations, such as *Melaleuca leucadendron* L., 9-20m in height, became shrubs hardly attaining 4m in higher elevations. But among those plants that were endemic to the mountaintops, a number seemed to have accommodated themselves completely to the great air movements.

One of them he described as a new genus: *Dracophyllum* Labill. "I have given this plant the name *Dracophyllum verticillatum* Labill., because its flowers are arranged in a ring. The species seems frequently to be unisexual. Its leave are coriaceous and denticulate, leaving a scar on the stalk when detached, as occurs in all species of *Dracaena* Vand. ex L. with which this plant has great affinity, including the texture of its wood. Therefore, it is within the Monocotyledons despite having both calyx and corolla, and it ranks naturally just following the Asparagus Family."[253]

From the mountaintops, they could see not only the immense line of reefs that protected New Caledonia on both coasts, but to the south a beautiful valley bordered by large plantations of coconuts. Substantial tracts of the lowest terrain appeared to be cultivated, suggesting not only a substantial population but a people different from those they had met. And, in fact, two distinct physical types would be recognized in the 19th century within the native population. One, Polynesian, with tall, well-proportioned body, the skin fairly light; the other, sub-Papuan, with smaller, thin-limbed body, black hair and dark skin. The latter type fitted the descriptions of the New Caledonians given by both Labillardière and Entrecasteaux.

Labillardière was obliged to remain on board all day the 26th to describe and prepare the collection of the previous day. His work was interrupted by the arrival of two aborigines who had swum out to the frigate. They made some effort to assure the French that they had not been part of the group that had committed

253. Labillardière. 1800. *Relation*. 2: 211. Tab. 40. He was right that the *Dracophyllum* are monocotyledons *in habit*. But the genus is recognized today within the Ericaceae.

hostile acts, indicating that they had eaten two natives shot in the recent skirmish. The French did not entirely believe the self-serving story.

Even so, they had brought a tool they called *nbouet*: a fine piece of flattened serpentine with cutting edges, nearly oval in shape, completely polished, and about 2dm long. Pierced by two holes through which two very flexible sticks had been threaded, the stone could be fastened to a wooden handle, bound tightly with strands of bat hair. The tool was used to cut up the bodies of enemies killed in battle. One of the natives gave a simulated demonstration of the procedure using one of the crewmen who agreed to lie down on his back.

The aborigine first gave a representation of a fight in which an enemy was felled by blows from a club and a spear. Then, after executing a sort of pyrrhic dance, he showed that they began by opening the victim's abdomen with their stone tool. The intestines were then removed and put aside using another tool fashioned from two well-polished human forearm bones. Following which the genital organs were removed to become the victor's share of the vanquished. The arms and legs were next severed at their joints to be distributed, along with other parts, to each of the participants in the fight for family use. The flesh of the unfortunate was then to be devoured after having been grilled over embers. The cannibal gave them to understand that the flesh on the arms and legs would be cut into slices of 7-8cm of thickness, and that the muscular portions were regarded to be the most desirable food.[254]

As there could no longer be any charitable doubts about local cannibalism, the naturalists did not venture into the interior until 28 April 1793 when a well-armed party had been assembled to assure safety. Their destination was the beautiful valley beyond the mountains they had perceived several days earlier. Once within that valley, Labillardière made an abundant collection, among them the fern *Acrostichum australe* L.f. [=*Actiniopteris australis* (L.f.) Link] and several new species of orchids.[255] A sudden downpour obliged them to seek the shelter of large rocks, finally forcing them to abandon their plan to spend the night in the field. On the return trip, Labillardière collected several species of passion-flower.[256] Ginger, *Amomum zingiber* L., was growing quite abundantly, but the natives told him they made no use of it.

254. Ibid. 2: 215-217. For a definitive commentary on the eno-Rousseauian denials that culinary canniablism ever exited anywhere, see Lawrence H. Keeley. 1996. *War Before Civilization* (New York & Oxford; Oxford University Press), 103-106. Useful distinctions are made between *ritual, culinary*, and *starvation* cannibalism.

255. *Limodorum unguiculatum* Labill. 1824. *Sertum Austro-Caledonicum* 1: 19-20. Tab. 25. [=*Calanthe striata* R. Br.].

256. *Disemma aurantia* Labill. 1825. Ibid. 2: 78. Tab. 79. [=*Passiflora aurantia* (Labill.) Forst. f.].

A second attempt was made to reach the cultivated areas in the valley across the mountains beginning on 2 May 1793. It seemed probable that they would encounter many aborigines; but the party, numbering 20, was well-armed and confident any attack could be repulsed. Following a more westward route initially on that outing, they came across an abandoned hut sometime before midday. Its neighboring plantations had been devastated only recently. One could see the remains of sugarcane and Caribbean cabbage (*Arum esculentum* L.), and all the coconut trees had been cut off toward their tops. They could only wonder whether the inhabitants had been cannibalized by their despoilers. After lunch, they came across additional evidence of devastation: a main dwelling destroyed and all the surrounding coconut trees beheaded.

Encountering marshes toward the end of the day, the party veered southward, soon finding a hill of difficult access as a secure place to spend the night. With daylight, 3 May 1793, they began the descent toward the cultivated plain, proceeding down a delightful dale where good soil and ample moisture promoted a vigorous growth of plants. Labillardière made a large collection in that area including a new species of fern.[257] Before reaching the plain, several members of the party, arguing that provisions would soon be exhausted, but possibly fearing an encounter with large bands of aborigines, elected to return to the ship. That reduced the party to 15, but they pressed on nevertheless.

Upon reaching the targeted plain, their first sight was another dwelling entirely destroyed and several neighboring coconut trees cut down at their base. Farther on, they came upon plantations of yams and sweet potatoes, but no aborigines were visible. Eventually they came upon an elderly man who was busy pulling up roots of the legume *Dolichos falcatus* Klein ex Willd., passing them to a child for cleaning. A second aborigine soon approached, and the two followed the French party to the edge of a saline channel that cut across the plain. At that point, the aborigines pointed to a trail that could take them back across the mountains.

They reached the summit shortly before darkness and camped for the night. On the morning of 4 May 1793, they consumed the last of their supplies and recognized the need to get back to the ships. But Labillardière could not resist his desire to spend several hours in a delightful grove a short distance from where they had spent the night. He collected a substantial number of plants he had not previously seen on New Caledonia, mostly within the Proteaceae and the Bignoniaceae. One of the most beautiful, a shrub in the Elaeocarpaceae, he thought to represent a new genus, *Antholoma*.[258]

257. *Marattia attenuata* Labill. 1824. Ibid. 1: 9-10. Tabs. 13-14.
258. *Antholoma montana* Labill. 1800. *Relation*. 2: 236. Tab. 41. [=*Sloanea montana* (Labill.) A. C. Smith.]

He used the entirety of 5 May 1793 to describe and prepare the ample collections made during the field trip, and Entrecasteaux noted that Labillardière seemed especially pleased by the results.[259] Their pleasure turned to sorrow the next morning when they learned that Captain Huon of the *Espérance* had died during the night from the consumptive or wasting fever from which he had suffered for several months, virtually confined to his cabin. Entrecasteaux, in truth, believed that Huon's health had been undermined by years of difficult service. Yet, toward the end, he had expressed not the slightest regret about having undertaken the search for La Pérouse. He had made known his desire that his body be put beyond the reach of the ferocious inhabitants of New Caledonia. Accordingly, the General thought it proper to have the burial on an off-shore island rather than on the mainland. The funeral was held at night to conceal it from the natives, and no marker was left for the same reason.

On 7 May 1793, in preparation for departure, the general reassigned Captain d'Auribeau to the command of the *Espérance* and gave the command of the *Recherche* to Lt. Rossel. No reliable traces of La Pérouse had been found; their major water source had been drying up; and the natives' eagerness to steal had led to several armed skirmishes. The less than amiable association with any of the natives had made it impossible to develop a local vocabulary as had been done on other islands. And, as on Tongatapu, they had seen evidence that tribal superiors could take anything they wanted from social inferiors.

That being the case, Entrecasteaux was perplexed by the disparity between Cook's more favorable description of the inhabitants and what the French had been observing. He wondered if perhaps the material artifacts from Europe, given benevolently to the natives by the Cook expedition in 1774, had aroused jealousies and fighting over who should possess them. Could it be that continual raids and killings over the subsequent 20 years had altered the character of the society? The substantial number of abandoned dwellings, the fields no longer cultivated, the excessive poverty and undernourishment were distressing signs of degeneration. But he also recognized a simpler explanation: that the peaceful population of Cook's day had been overrun and replaced by a more warlike tribe, leading to the decline of agriculture and the appearance of cannibalism.

As they had not found any descendents of the pigs Cook had left at that anchorage, there seemed to be no point to leaving livestock on the island. A fierce people, who did not even spare their own kind, were unlikely to give domestic animals time to reproduce. Had Entrecasteaux not been misled by Cook's obser-

259. Rossel. 1808. *Voyage d'Entrecasteaux*. 1: 340-341.

vations, he concluded too late that his visit to New Caledonia could have been avoided. Given the actual brutality of the aborigines there, it would have been impossible to conceive that a shipwrecked party seeking asylum could have met with anything but death.[260]

Labillardière, in a party of 6, did make one last, well-armed foray into the forest to the south-southeast on 8 May 1793, finding plants not previously collected. He encountered a native breadfruit tree, only the second one he had seen on the island, both of them having only male flowers. (The genus is considered to be monoecious.) Their leaves seemed less deeply divided than those on Tongatapu, so he took 3 cuttings for preservation in the storage case.[261]

There had been ample opportunity to make abundant natural collections on that northern part of New Caledonia, but otherwise the visit had been regrettable. Beyond the failure to have established an informative exchange with the aborigines was the death of Captain Huon. Entrecasteaux was deeply smitten by the loss of his longtime friend and associate, a grief made worse by awareness that he had encouraged Huon to join the expedition in the knowledge of his questionable health. When the order was given to set sail early on the morning of 9 May 1793, a change of scene was clearly overdue.

260. Ibid. 1: 343-359.
261. Labillardière. 1800. *Relation*. 2: 244-248.

XIII
The Voyage of Entrecasteaux: Part 7
Its Woeful End

During the ensuing weeks, as the two frigates passed through various island groups, the occasions for botanizing were no more. All attention became concentrated on the search for La Pérouse. After exploring the reefs northwest of New Caledonia for four days, Entrecasteaux broke away on 13 May 1793 and set a course for Santa Cruz Island, also called Nendo Island, the largest in the Santa Cruz group.

That group came into view on 19 May 1793. They sighted several of the smaller islands that had been charted by Philippe Carteret in 1767, now Anuta and Utupus Islands. And they took a bearing on a third island that evidently had not been perceived by Carteret and named it Ile de la Recherche (now Vanikola Island). They saw it as such a great distance that they could not put it precisely on their maps.

The atlas later published by Rossel provided a comparison between Carteret's maps of the Santa Cruz Islands and the one prepared by Beautemps-Beaupré.[262] Neither correlate well with a contemporary map of that region, permitting some uncertainty about Entrecasteaux's exact route through that archipelago. That imprecision concerns us only for the sad fact that the expedition passed close to the Vanikola Island where evidence of La Pérouse's shipwreck would ultimately be found and named it the *Recherche*.

Proceeding directly toward the east coast of Santa Cruz Island, Entrecasteaux sought a safe anchorage from which he could seek information about La Pérouse. On 21 May 1793, he sent out several small boats under the command of Ensign Willaumez to take soundings, while many native pirogues came close to the frigates to engage in trading. Entrecasteaux found the natives unattractive, sinister in appearance, and even more inclined to theft than the aborigines of New Caledonia. They brought out two species of breadfruit, both seedless, as well as yams, coconuts, and sweet potatoes to trade.

As for Willaumez, the coastal indentation he had been sent to probe as a possible bay for an anchorage turned out to be only a deep-water channel between two islands. He had undertaken some peaceful trading with natives who had surrounded his boats, when suddenly a native in a pirogue stood up with a bow. He

262. Rossel. 1807. *Voyage d'Entrecasteaux*. 3. Tabs. 19-20.

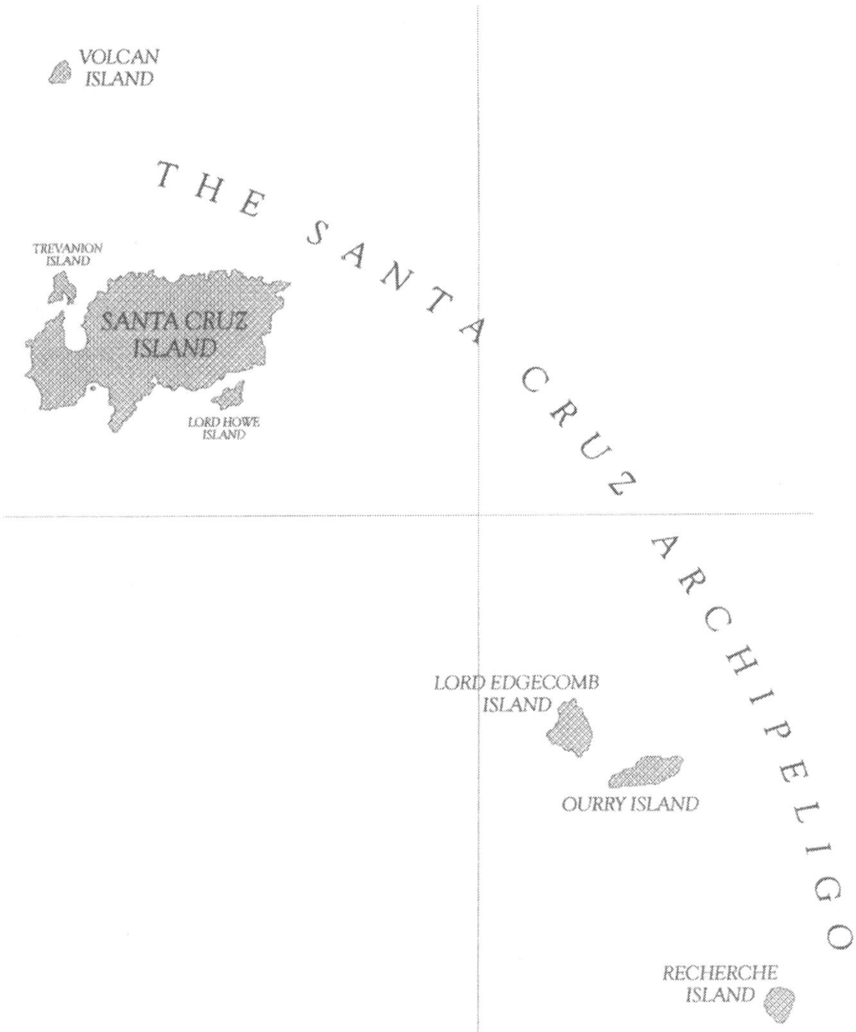

Modern longitude and latitude cordinates of represented area are from 165°38′ E. to 168° E. and from 8°31′ S. to 11°40′ S.

let an arrow fly. Even though the distance was more than 80m, the arrow struck the forehead of one of the crewmen. Both French boats returned the fire. The native was seen to fall, his companions leaped into the water, and the remainder of the pirogues fled to shore. The abandoned pirogue, with its bows and arrows, was towed back to the *Recherche*.

The sailor who had been struck by the arrow had only a superficial wound and felt only slight pain. But the French, suspecting that the natives poisoned their arrows, carried out experiments with the captured arrows. Several of their chickens were slightly pierced with arrow tips, but exhibited not the slightest negative symptoms. Moreover, the sailor's wound healed quickly. Poison had to be ruled out. Yet, two weeks later, he developed a violent fever and died after three days. Labillardière diagnosed the fatal illness as tetanus.[263]

As further association with that group of natives seemed unlikely to be productive, the General gave instructions, as soon as Willaumez's boats had been hoisted aboard, to sail along the coast in search of a more favorable reception. An attractive bay along the north coast seemed to be a good place to anchor and an opportunity to see if there was any evidence of European equipment or clothing. The swell proved to be too great for the longboats to risk a landing. A limited trade with natives in pirogues took place, coconuts for red cloth; but the natives, armed with bows, seemed mistrustful, and Entrecasteaux thought it best to move on to avoid a repetition of the earlier incident. In leaving Santa Cruz Island for the Solomons, Entrecasteaux lost his last possible opportunity to gain information about the La Pérouse site on Vanikola Island.

Once in the Solomons, apparently along the eastern shore of San Christobal Island, they experienced a new kind of native behavior on 28 May 1793. Aborigines would come out to the frigates in pirogues with little or nothing to trade, but would make elaborate signs to invite the visitors to land. As the Solomons had been named the Land of the Assassins by an earlier French explorer, the intent behind such invitations was necessarily questionable. The enticements failing, about a dozen arrows were shot from an unusually large pirogue at crewmen on board the *Espérance* wounding one seaman in the arm. One shot was fired at the pirogue in return, hitting no one. But as some rockets were then being prepared to send signals, one of them was launched into that pirogue. It cannot have caused any great harm, but it provoked the desired panic. Suspecting the natives of complete perfidy, Entrecasteaux ignored all further invitations to land.

Moving ahead along the route projected by La Pérouse, the frigates approached Guadalcanal Island on 30 May 1793 and sailed along its south coast the following day, reconnoitering the coast but not landing in the bad weather. Instead, they turned westward to reconnoiter the Louisiade Archipelago. On 11 June they approached what is still known as Rossel Island, but could not move in close because of an immense number of shoals. Navigation in the region proved to be exceptionally treacherous: reefs, shoals, and strong currents, giving little room to tack.

263. Labillardière. 1800. *Relation.* 2: 251-256; Rossel. 1808. *Voyage d'Entrecasteaux.* 1: 362-371, 401-403.

During the next weeks, the voyage became increasingly frustrating. Not only was the navigation difficult, but subsequent encounters with natives, who threw stones and even spears at personnel on board, left no doubt that they were hostile. By the time they reached New Britain Island, 29 June 1793, both conditions and supplies on board were deteriorating seriously. The crews were fatigued and health generally affected. While reconnoitering New Britain, they failed to find any cultivated areas; meaning that even if a suitable landing place could be found, it was unlikely to be a source of supplies. Recognizing that a diet of wormy hardtack and salted meat could not long sustain them, and that even coffee had had to be abandoned as it was contributing to highly inconvenient intestinal cramps, the General reached the decision to break off the campaign. On 8 July 1793, he gave orders to sail directly for Java where the Dutch could offer great resources. It was to be his final journal entry.[264]

Lt. Rossel, then serving as captain of the *Recherche*, assumed responsibility for the journal on 9 July 1793, a clear sign that Entrecasteaux was ill. Scurvy had become increasingly apparent amongst crewmen, and the General was also afflicted. Entrecasteaux had favored reaching a Dutch base, whether Java or in the Moluccas, by sailing along the north side of the Admiralty Islands to avoid the calms experienced closer to New Guinea in 1792. But, by 17 July 1793, the senior officers, alarmed by the General's deterioration, proposed to go ashore on Waigeo Island just northwest of New Guinea for recuperation; and to get there as quickly as possible by allowing the *Recherche* to separate from the slower *Espérance*. Entrecasteaux opposed that proposal, arguing that he would not violate his instructions for personal advantage.

Only on the evening of 19 July 1793 did he give way, permitting the two ships to proceed separately to Waigeo Island. By daybreak on the 20th of July, the *Espérance* was barely visible in the distance. At which point, Renard, chief-surgeon, announced that he had exhausted his knowledge in attempting to relieve the General, saying he needed to consult Joanet, chief-surgeon on the *Espérance*.

We know from Labillardière and others that the malady had been diagnosed as a violent or spasmodic colic, and may have been a peptic ulcer.[265] So, the *Recherche* had to maneuver to allow the other frigate time to rejoin. The two surgeons then tried to relieve the General's pains by putting him into a bath, but that only brought on terrible convulsions and a complete loss of consciousness. The two surgeons, aided by Labillardière, were unable to halt the spasms, and Entrecasteaux

264. Rossel. 1808. *Voyage d'Entrecasteaux*. 1: 386-392, 422, 434-439; Labillardière. 1800. *Relation*. 2: 257-286.
265. Labillardière. 1800. *Relation*. 2: 287; Horner. 1995. *Looking for La Pérouse*, pp. 189-190.

died at 7:30 on the evening of 20 July 1793 without having regained consciousness.[266]

Religious and military services were held the following day with burial at sea, and the normal solemnity of such an occasion was deepened by a general recognition of the magnitude of the loss. Rossel strongly implied that none of the remaining naval personnel came close to Entrecasteaux in his intellectual grasp of the various technical knowledges critical to the expedition. It simply could not go on without him. Captain d'Auribeau, as senior in rank and age, immediately became chief of the expedition, but he inspired none of the confidence nor the affection that Entrecasteaux had enjoyed. His own health, moreover, had been undermined and imperiled his ability to act decisively. The subsequent two-week void in Labillardière's journal may well be a measure of the consternation on board the frigates.

The continuing voyage toward a Dutch base was maddeningly slow because of feeble wind, and the increasing number of cases of scurvy made Auribeau steer for an anchorage on Waigeo Island. That island was first sighted on 11 August 1793, but valuable time was lost in fruitless reconnoitering. Ultimately, a native in a pirogue, evidently a Malay, proved willing to help once he understood the visiters were not Dutch. He recommended an anchorage on Boni Bay on the northeast coast of Weigeo. There they anchored on 18 August 1793.

All the sick were immediately taken to shore where they almost at once experienced major swelling in all parts below the waist. The affliction was very frightening, but the chief-surgeon gave assurances that it would only be temporary. After remaining seated for about six hours, the men felt sufficiently revived to walk easily to the boats that took them back to their ships.

Because the source of fresh water was some distance from shore, it took nearly nine days to refill the ships' barrels and cut sufficient wood. Each day, native pirogues brought out fresh produce for trade: large turtles that could be turned into a bouillon that provided nourishment for the sick; also cooked turtle eggs, smoked turtle meat, chickens, pigs, oranges, grapefruit, coconuts, papayas, gourds, purslane (*Portulaca quadrifida* L.), sugarcane, yams, sweet potatoes, bananas, lemons, and pimentos.[267]

The sojourn gave the naturalists a good opportunity for fieldwork. Labillardière not only made a rich collection of plants, but shot many rare birds for the collection. Captain d'Auribeau took advantage of the sojourn to have his headquarters moved to the *Recherche*, moving Rossel to the *Espérance*. The adjustment

266. Rossel. 1808. *Voyage d'Entrecasteaux*. 1: 440-445.
267. Labillardière. 1800. *Relation*. 2: 288-289; Rossel. 1808. *Voyage d'Entrecasteaux*. 1: 451-457.

may have symbolized Auribeau's position as commander-in-chief, but he knew himself to be unequal to the command. He had been sufficiently ill at the time of Entrecasteaux's death to excuse himself from attendance at the funeral. And now, just before sailing from Waigeo on 27 August 1793, Auribeau told Rossel that he had been in bad health for several days and gave Rossel authority to direct the operation until he had recovered.

On the day after departure, the *Recherche* signaled that Joanet, surgeon on the *Espérance*, was needed on board. Joanet found Auribeau in serious condition. He had experienced convulsions during the night and remained unconscious. Once he was informed, Rossel thought it best to steer for the port of Kajeli on Buru Island in the Moluccas, closer than Surabaya on Java, in the hope of treating Auribeau and giving the entire crew additional opportunity to fortify themselves for the long voyage home.[268]

The ships reached Buru Island on 3 September 1793 and anchored off Kajeli. The Dutch Resident, quite aware of navigators' needs, gave permission to anchor and sent out pirogues bearing fish, fruit, yams, and freshly-killed buffalo meat. It is evident that the Resident had not as yet had any word that France had declared war against Britain, Holland, and Spain on 1 February 1793, shortly after the execution of Louis XVI. Nor could Rossel have known that fact, the expedition not having been in a European port since October of 1792.

Lt. Rossel would now make some observations that revealed a considerable change in outlook since the previous year when Entrecasteaux and Labillardière, when visiting Dutch outposts, had registered their disapproval. Rossel, as a naval officer, was not counted among the naturalists. Yet, after the departure of the troublesome abbé Bertrand at the Cape of Good Hope, Rossel had been exercising his duties as chief astronomer on the *Recherche*. The astronomical detail he would publish as the second volume of the *Voyage* attests both to his conscientiousness and his scientific knowledge.[269] He was no mere nuts-and-bolts seaman.

"At the beginning of the campaign," Rossel wrote, we had taken as much pleasure in gazing upon the beauties of primitive Nature in new countries as we now did in rejoining a cultivated land and civilized men. The very beauties of raw Nature, which had at first enraptured us, now struck us as a dismal monotony. We had begun to feel only distaste when encountering forsaken lands such as Australia. That sense of genuine curiosity, which had aroused the desire in us to visit primitive peoples and to learn about their customs, had become entirely extinct. Such men, so close to the state of Nature, and about whose simple [goodness] we

268. Rossel. 1808. *Voyage d'Entrecasteaux*. 1: 463-467.
269. Horner. 1995. *Looking for La Pérouse*, p. 41.

have had such exaggerated ideas, now only inspire painful memories in us. We have seen some among them commit the most outrageous excesses of barbarity, and all of them were even more corrupted than civilized peoples. Our eyes, long since wearied by the spectacle of arid and desert coasts, have now fallen with quiet satisfaction upon a fertile country, reminding us of our traditional ways of life. And our soul, previously burdened by thoughts about those fierce peoples, was relieved by the appearance of the town of Kajeli, its mosques and its houses numerous enough to constitute a kind of city. We could wish for no more than to come closer to our *patrie*. Given this separation from the land of our birth, every European becomes a compatriot, every Frenchman a member of our own family."[270]

This favorable impression was reinforced by the cordiality of the Dutch Resident. He devoted himself enthusiastically to furnishing all the supplies needed on board the ships. He provided quarters on shore for Auribeau and for the naturalists to facilitate their field trips; and a large house was made available for the sick. The simple, calm life was enjoyable after so much hardship, a life that, under any other circumstances, the French would have found monotonous in its uniformity. But after a study of the local government and economy, even Rossel could not avoid observing that all regulations by the Dutch company were meant to prevent the native Malays from being prosperous, a condition that might encourage a native industriousness that could threaten the company monopoly.[271]

Labillardière did spend his days exploring the island, but his journal entries became sparse, not fully reflecting his observations. He noted that the sago palm (*Metroxylon*) was widely cultivated in swampy areas and provided the natives' principal food. *Melaleuca latifolia* Raeusch. grew abundantly on hillsides, and he was shown how cajeput oil was obtained from its leaves. He was told that teak (*Tectona grandis* L.f.) was grown beyond the town to a height of 40m and used by the Dutch in the Moluccas for shipbuilding. One gets the sense from his journal of a winding down, a need for rest. Rossel claimed that everyone benefited from the sojourn except for Auribeau, who did not improve.

They set sail on 15 September 1793, Surabaya on Java still being the goal. As he approached Butung Island, Rossel's route to the open sea was closed off by contrary winds and adverse currents, forcing a decision to pass through an inner channel between Celebes and Mounan Island. Toward that end, he sent out a boat on 24 September 1793 with Beautemps-Beaupré and Legrand to explore the coast of Mounan Island to locate an open passage, a two-day venture.

270. Rossel. 1808. *Voyage d'Entrecasteaux*. 1: 470-471.
271. Ibid. 1: 472-478.

Labillardière took advantage of the delay to go ashore on Butung Island and, after a very difficult landing, pushed inland. He found many plants he had not previously seen: a nutmeg tree, *Myrisica uviforma* Lam., whose fruit was not aromatic; *Cynometra ramiflora* L., a legume; and various species of rattan palms (*Calamus* L.) that sometimes produced stalks several hundred meters in length. *Bombax ceiba* L., the cotton-silk, and several other species of the same genus, bore fruit that provided abundant food for numerous bands of monkeys.

Subsequently, Labillardière landed on Mounan Island. The total absence of a native population caused Labillarière to speculate that the swampy ground on both islands had probably been avoided as unhealthful. Because an extremely contagious dysentery had begun to afflict all personnel on the frigates, almost as a noxious substitute for scurvy, he presumed that the germ of the disease might have been acquired from the two islands. Yet, he also knew that on Buru Island there were equally dangerous swamps near the town. Whatever the source, the personnel had begun to suffer a debilitating and dangerous disease.[272]

These explorations ended by 29 September 1793 when the inner channel had been deemed too obstructed by reefs and shoals to proceed safely. Accordingly, Rossel took the strait between Butung and Mounan Islands. They emerged on 5 October 1793 to find a good anchorage on the coast of Butung. A great many pirogues came to meet them bringing fowl, yams, and pumpkins. They also traded monkeys, parrots, and cockatoos to the seamen in exchange for old clothes. Virtually every crewman bought a pet. Labillardière again went ashore where he collected *Barleria prionitis* L. and what he thought were several new species of *Croton* L.

Another day of sailing brought them to the town of Butung (Baubau), 8 October 1793, where Rossel meant to take on sufficient supplies to get to Surabaya. This was the domain of the Sultan of Butung, as the Dutch garrison had been massacred more than a century earlier. The local population feared, not a return of the Dutch, but Papuan pirates who frequently raided the region and carried off the young of both sexes for slaves. But it was also evident that the local Malays themselves had slaves for sale, prisoners taken in warfare.

It was imperative to hurry. Though the supplies purchased were of good quality, they were not the sort that preserved well. Moreover, dysentery was making rapid progress on both frigates, and the heat had become very great. The considerable number not fit for duty had become alarming. When they left Butung on the afternoon of 9 October 1793, Auribeau had again assumed his command.

272. Ibid. 1: 482-488; Labillardière. 1800. *Relation*. 2: 298-302.

The sight of Madura Island off the entrance to Surabaya on 16-17 October 1793 promised relief so desperately needed.[273]

On the morning of 19 October 1793, Lt. Trobriand, the second-ranking officer on the *Espérance*, took a longboat into Surabaya to announce the arrival of the expedition to the Dutch governor: to tell him the needs of the expedition, and to ask permission to enter port. He did not return within the expected time, and by the 23rd the French became extremely anxious. The Dutch were known to protect all who sought their assistance. Yet, it had been noticed that the local fishing boats had kept their distance, suggesting the possibility that the silence from Surabaya meant that France was then at war with the Dutch. Even so, the French could hope to be regarded as neutrals in that they were on a campaign to increase mankind's knowledge, which had nothing to do with the quarrels between nations.

The alternative, the possibility that they would be forbidden to enter port, was frightful. They were hardly in any condition to sail on without putting the one-third of the crew still healthy at risk, thus putting the fruits of the expedition and the safety of the frigates in doubt. Food and water supplies were low, and Mauritius was at best a six-weeks voyage away. On 23 October, therefore, Auribeau sent a second boat from the *Recherche* with Ensign Mérite as negotiator to ask for the courtesy of pilots to guide them to an anchorage, and sent him under a white flag.

On 25 October 1793, the Dutch had a letter from Lt. Trobriand brought out to Auribeau. In it, he had to announce not only that he was a prisoner of war, but the stunning news that France was at war with much of Europe, that the monarchy had been overthrown, and that a republic had been proclaimed. It is not difficult to understand that for men whose political sensibilities dated from 1791 such a turn of events by late 1793 was either dismaying or exhilarating. From that moment forward, no decision that had to be taken under the appalling circumstances in which the expedition found itself could be free from the suspicion that a political motive underlay it. When the requested pilots were provided and the frigates anchored at Surabaya on 27 October 1793, the scientific mission and the search for La Pérouse had come to an end. Lt. Rossel terminated the Entrecasteaux narrative.[274]

273. Rossel. 1808. *Voyage d'Entrecasteaux*. 1: 483-514.
274. Ibid. 1: 515-520.

XIV
Internment

Another narrative, frequently cited by others for the history of the Entrecasteaux expedition, that of Pierre-Roch Jurien, was only set down after his retirement as a vice-admiral about 1825. Precise dating by then proved to be impossible. Consequently, the most valuable part of the young ensign's observations proved to be his statements about the political climate on board. As a prospective naval career officer, his sympathies were with the naval officers, not with the scientists. (Jurien was from a bourgeois family, ennobled only in 1815 as vicomte by Louis XVIII.)

Jurien, in all fairness, was very young when he served on the *Espérance* and gave little evidence in his account that he had really understood the scientific purposes of the expedition at the time; and his later professional advancement, depending upon royal favor, did not contribute to objectivity. While he characterized the naval officers as having been a carefully-selected elite who knew their jobs very well, he added that unfortunately there had also been a group of scientists on board, a group both distinct and potentially uncongenial. He saw them as excess baggage and the source of hostilities and frictions.

After the death of Entrecasteaux when the command had devolved upon Captain d'Auribeau, Jurien recalled that "The Revolution of '89 had just inaugurated the reign of the Encyclopédistes. Our commander had no sympathy for the new tendencies. He had been contending in good humor with the invasion of his ship by Science, which, despite him, had been installing breadfruit trees right up to the poop. All these alleged benefactors of humanity, with herbaria meant to endow France, with kitchen-garden seeds meant to enrich desert beaches, reminded me of the well-known play by Shakespeare, *Much Ado About Nothing*."[275]]

Frank Horner's account of the immensely confusing story of Auribeau's conduct and the internment of the French by the Dutch is as balanced a treatment as we are ever likely to have. Its very complexity amounts to a prediction that the entire truth of that situation is unlikely ever to be known. The arrival of two French warships in time of war had to be alarming, no matter that they were lightly armed and hardly formidable. Moreover, in the best of times, the Dutch

275. Pierre-Roch Jurien de la Gravière. 1860. *Souvenirs d'un amiral.* Paris: Hachette. 1: 85, 141, 227-228.

company officers had always been wary of foreigners, determined to protect their monopoly on the trade in spices.

The local Dutch governor believed that the state of war annuled the safe-conduct guarantee under which the expedition had sailed. As a consequence, he ruled that if the ships entered the port, all arms and both rudders would have to be turned over to the Dutch; and he meant to defer a decision about the status of the French to the High Regency in Batavia (Jakarta). The most controversial among the governor's conditions was his demand that all journals and personal records of the voyage be turned over to the Dutch in order to ascertain that they contained no information threatening to Dutch security.

Auribeau's conduct in the crisis became controversial immediately and ever after. Did he bend to the governor's demands because the alternative was so bleak? Was his response that of a sick man who lacked the vigor for the risky alternative? As a royalist, did he privately perceive the Dutch as an ally against the revolutionary government in Paris? And, accordingly, did he anticipate a benign ruling from the High Regency?

There remains clear evidence that most of the French officers preferred the danger of a dash for Mauritius than to accept Dutch conditions. Some, like Jurien, believed that Entrecasteaux would have opted for that course.[276] But the most divisive of Auribeau's decisions was his order to have all personal journals and records turned over to him for Dutch examination. He could argue that the order was consistent with his instructions to have all such material turned over to the commander-in-chief at the end of the expedition, and that the expedition had effectively come to an end. The refusal to comply with the order by some members of the French staff bred immense distrust.

Labillardière, for instance, valued his journal as a botanical record essential to his eventual publication of new species and was determined to conceal it. Moreover, a year earlier, at Ambon, he had written a description of its crumbling fortress and its small army, men in virtual servitude and of questionable reliability, which surely implied that the Dutch would be easy prey to a European invader.[277] If such a description fell under Dutch eyes, any hope of immediate repatriation would vanish. If one recalls, finally, that during that sojourn on Ambon rumors had circulated that more radical members of the staff were collaborating on a "republican" account of the voyage for private publication, such members can only have suspected that Auribeau had seized upon an opportunity to confiscate documents he feared to be offensive.

276. Jurien. 1860. *Souvenirs.* 1: 256.
277. Labillardière. 1800. *Relation.* 1: 346-347.

On 30 October 1793, several days after Auribeau's acceptance of Dutch terms, the French were given permission to reside in Surabaya. Ten days later, the Dutch had a change of mind and revoked that permission. All but the sick had to return to the frigates. It soon became evident that such captivity was intolerable. The number of sick kept augmenting, either from dysentery or some malignant fever, until half the personnel on board was afflicted and the deaths were continuing. The Dutch were obliged to reverse themselves again, and the French were allowed to go ashore.

On 11 December 1793, the governor granted the naturalists the liberty to visit the Prau (Prahu) Mountains to the west-southwest of Surabaya. They departed the next day with native Javans to carry their effects. Labillardière was particularly pleased to find species within *Uvaria* L., *Helicteres* L., and *Bauhinia* L. By December 15, they had moved on into the Panangounan Mountains. There they encountered substantial swamply areas covered by the very large leaves of the Egyptian lotus, *Nymphea nelumbo* L. [=*Nelumbo nucifera* Gaertn.].

Returning to Surabaya on 19 December 1793, Labillardière and Riche made plans to botanize in the Passervan Mountains. Auribeau, whose responsibility it was to apply for the permit, was unable to obtain it, explaining that word had come from Batavia that the naturalists were not to be allowed to go farther than a 3 to 4-hour walk from town. Consequently, Labillardière and Riche, who lived in the same house, carried on their research in a manner that enabled them to return to Surabaya every evening. The arrangement lasted for about two months until the Dutch had yet another change of mind.

Frank Horner has deduced quite reasonably that Dutch mistrust of the French as the enemy in wartime had developed new dimensions as the weeks wore on. It seems evident that Auribeau, for reasons of his own, had seen fit to explain the political divisions within the French camp to his Dutch captors during his negotiations with them. His implication would have been that the Dutch were at war, in fact, with French republicans, not the Bourbon loyalists. This led the Dutch to fear a republican mutiny against Auribeau, a seizure of the two frigates, and an attempt to flee to Mauritius.[278]

This will account for the sudden occupation of the frigates by Dutch troops on 19 February 1794. The French lodgings on shore were surrounded, and the people indicated by Auribeau to be republican were arrested for close detention: Ensign Legrand, Ensign Willaumez, Lt. Laignel, Claude Riche, Louis Ventenat, Piron, and Labillardière. Two domestics from the *Espérance*, Michel Sirot and Pierre Creno, chose to remain with them. Jurien and Beautemps-Beaupré were

278. Horner. 1995. *Looking for La Pérouse*, pp. 203-211.

held momentarily, but released after pleading political neutrality. In the aftermath, Auribeau had the Bourbon flag hoisted on his ship, further suggesting his common cause with the Dutch.[279]

Those arrested were taken to Samarang for detention, so that Labillardière was separated from his collections. It is a curious fact that, from that moment on, he would always refer to the collections as *his*, whereas it had been understood from the beginning of the expedition that all collections would be delivered to the king. As he offered no rationale to explain his assumption of ownership, it may well be that, with the fall of the monarchy, Labillardière had simply assumed that the crown had lost its property rights. Being far removed from events in France, he could not know that all crown properties had been nationalized by the Republic. It is also quite possible that he had assumed from the start that, even though all collections were to be delivered to the king, they would thereafter be returned to their collectors as the logical classifiers and publishers of the new information.

As for Félex Delahaye, the gardener, he had never been regarded as a scientist by the naturalists, merely as someone who knew how to plant, sow, and irrigate, a distinction roughly between gentlman and servant. Auribeau's list had designated only gentlemen for detention, so that Delahaye remained free to move about within a designated sector of Surabaya. He kept his notes, his collections of seeds; and, in a letter to André Thouin from Java, dated only 1794, he reported that he maintained 30 roots of very vigorous breadfruit trees brought from the Tonga Islands. He expected to remain in Surabaya until peace could be restored and then take his trees to Mauritius.[280]

The Dutch collaboration with Auribeau in arresting his republican colleagues had an ulterior motive. The company authorities had recognized that the only way they could recover the expenses required to maintain their unwelcome guests would be to seize the two frigates and sell them. Auribeau's attempt to preserve the ships for the royal cause by raising the Bourbon standard proved to be futile. At the time of his sudden death from dysentery on 22 August 1794, Auribeau had been negotiating their sale with the Dutch; and the contract for the sale of the ships would be signed on 26 September 1794 by Lt. Rossel, his successor as commander.[281]

The seven republican prisoners, meanwhile, had arrived in Samarang on 11 March 1794 after a difficult overland journey. They found themselves confined to a hospital in a room with seven beds but neither chairs nor table. As they were

279. Labillardière. 1800. *Relation*. 2: 307-320.
280. Letouzey. 1989. *Le Jardin des Plantes*, p. 241.
281. Horner. 1995. *Looking for La Pérouse*, pp. 221-222.

neither ill nor wished to become so, it was immediately essential to appeal to the governor: to get him to sense the harshness of such an arrangement for men who, "on their return from a long and difficult voyage, undertaken for the progress of the sciences and the arts, believed they merited a different reception among a civilized people."[282] The appeal was successful, and, after several hours of necessary negotiation, the party was housed near the center of town and allowed to roam within a kilometer of the town.

On 5 April 1794, they learned that a Dutch ship was being prepared to sail within a few weeks from Batavia to Europe. Hoping to secure passage, they petitioned the governor for the right to send a delegation of two to Batavia. When the governor agreed, they drew lots to see which of the seven should enjoy the privilege of that opportunity. Riche and Legrand were the winners and left for Batavia the following day. Although the opportunity to go with that ship was denied, within two weeks the remainder of the party was invited by the governor to proceed to Batavia to await the next sailing.

Willaumez, Laignel, and Ventenat accepted the opportunity. As the annual Dutch convoy from Batavia to Europe traditionally sailed only after the onset of the northeast monsoon, six or seven months ahead; and as Batavia was reputed to be a place especially unhealthful for Europeans, Labillardière and Piron elected to remain in Samarang until the Dutch fleet would be ready to sail. As a result, the two missed an unanticipated opportunity to leave Java on 18 July 1794 when an arrangement for a prisoner exchange enabled the five republicans (and 19 crewmen from the expedition) to sail for Mauritius.

Captain d'Auribeau, still alive that July, had come to Samarang to complete negotiations for the sale of his two frigates. He was still intent upon seizing Labillardière's journal and induced the governor to order a search of Labillardière's effects. The search was carried out on 28 July 1794 and failed to produce the document. It does not appear that Labillardière ever revealed how he had managed to conceal his journal so successfully on several occasions, even managing to continue his entries. Auribeau died shortly after, on 22 August 1794. A week later, Piron and Labillardière moved on to Batavia to be housed in Fort Anké. The site, about three miles west of Batavia and surrounded on all sides by marshes, was insalubrious, but less so than the town itself.

The surface of the Anké marshes was covered by various plants growing close together, resembling a beautiful prairie in full vegetation. Great numbers of grasses, sedges, and lotus grew from the bottom of the stagnant waters. The intervals between them were filled by great quantities of *Pistia stratiotes* L., a free-

282. Labillardière. 1800. *Relation*. 2: 321-322.

floating species in the Araceae, which sustains itself on the surface of the water by means of sympodia. Labillardière believed that the *Pistia* absorbed a great part of the harmful miasmas.[283]

During the sojourn of seven months at Fort Anké, Labillardière found little of note to record in his journal. It had been, after all, the record of his observations and collections. Once he could not roam freely, he fell silent. With some reason, he attributed his captivity more to the French royalists than to the Dutch, and he now had ample time to cultivate his suspicions and grievances, especially as his collections were in the hands of his enemies.

At some point at Fort Anké, undated in the journal, the officer in charge of the fortress, who visited the prisoners regularly, mentioned the death of Louis Girardin to them. Medical examination had revealed him to be a woman, Louise Girardin. She had embarked on the *Recherche* at Brest in 1791 as a steward in the commissary. Stewards, like naval officers, did not undergo physical examinations before boarding ships, and she had escaped detection. She would have been about 40, looked much younger, and had all the features of a man. Labillardière, however, claimed that the truth had been suspected from the outset of the voyage. It would come to light that she had left a very young child behind in France after having been abandoned by a faithless lover, and was fleeing to escape from family wrath. It is probable that Captain Huot knew something of her story and had in some way been associated with her plan to enlist. If so, it is also probable that Entrecasteaux had been privately informed. She was publicly revealed only in death.

Her story recalled a prior deception. Philibert Commerson, the botanist accompanying Bougainville, had had in his service a very young, unbearded man as an aide and preparer of collections. The men on board had detected nothing out of the ordinary. Only during their sojourn on Tahiti was she perceived by the islanders to be a woman. She remained with Commerson on Mauritius in 1768 and was with him until his death in 1773, afterwards marrying a European soldier living on the island. In all probability she was the elusive Jossigny who signed the illustrations of Commerson's plants.[284]

The arrival of the northeast monsoon enabled the Dutch convoy to sail from Batavia on 5 December 1794. The convoy carried 32 members of the Entrecasteaux expedition, including Lt. Rossel, as well as the valuable documents and col-

283. Ibid. 2: 328-330.

284. Auguste Chevalier. 1953b. "Un grand voyageur naturaliste normand, J. - J. La Billardière (1755-1834)." *Revue internationale de botanique appliquée et d'agriculture tropicale* 32: 117-118. The sentimental version of Louise Girardin's death given in Jurien's *Souvenirs* was either a fabrication or evidence of a faulty memory.

lections, among them the journals of Riche and Ventenat. No explanation survives as to why Labillardière and Piron were left behind, allowing the conjecture that Labillardière was being disciplined for the concealment of his journal.

Prolonging his separation from his collections could only heighten his suspicion that a deliberate plan had been afoot to deprive him of them permanently. When he would later learn that Rossel, needing to reduce the size of the collections for reasons of space on the Dutch ships, had put Deschamps in charge of deciding what was to be left behind, that suspicion turned into the conviction that Deschamps had been given the opportunity to divert some of his colleagues' botanical material for his private use. As Labillardière had long despised Deschamps for his dubious professional skills and his royalism, such a conviction came easily. The validity of Labillardière's charges, unfortunately, can neither be proved nor disproved.

Indeed, one cannot even account for why Deschamps was selected for the expedition in the first place. He had been born on 22 August 1765 in St. Omer (Artois, now Pas de Calais) and had completed his work for the medical doctorate in 1788. That guarantees that he had had some botanical training, but there remains no record of significant botanical activity thereafter to account for his appointment in 1791; and he exhibited little zeal for exploration while on the expedition. Finally, when it came time to return to Europe in December of 1794, he arranged to remain on Java, ostensibly to study natural history.

He would remain on Java until 1802 before sailing for Europe on a Dutch ship. In the English Channel, the ship was captured by the British and its cargo deposited at India House as wartime booty. Deschamps' manuscripts were subsequently purchased and donated to the British Museum, from which it would later be possible to publish his itinerary that strongly indicated his intention to publish a flora of Java. The disposition of his herbarium has remained obscure, meaning that all his plant specimens were lost. In later years, Deschamps resumed a medical practice in St. Omer where he would die in 1842.[285]

As for other naturalists on the expedition, Louis Ventenat died of "Java fever" three days after reaching Mauritius in 1794. He was only 29. Claude Riche, long in poor health, chose to remain on Mauritius for several years in order to enjoy more leisurely botanizing. As his health continued to worsen, he returned to France where he died in 1797 at the age of 35. The gardener, Félix Delahaye,

285. Steenis, C. G. G. J. van, J. J. van Steenis-Kruseman, and C. A. Backer. 1954. "Louis-Auguste Deschamps, a prominent but ill-fated early explorer of the flora of Java, 1793-1798." ***Bulletin of the British Museum (Natural History) Historical Series*** 1, no. 2: 51-54. (These authors seemingly knew nothing about controversies surrounding Deschamps.) The war between France and Holland had ended with the Treaty of the Hague, 16 May 1795, forcing the Dutch into an alliance against Britain.

continued to tend his breadfruit trees on Java, along with those entrusted to him by Labillardière, until January of 1797, when he embarked for Mauritius. Some of the breadfruit he delivered were acclimated later in the French West Indies, and a few were sent to the hothouses of the Jardin des plantes in Paris.[286]

On sailing from Mauritius in May of 1797, Delahaye took a large collection of dried specimens and seeds destined for the Muséum de l'Histoire Naturelle in Paris. It is presumed today that his herbarium only contained duplicates of Labillardière's and Ventenat's plants. After arriving at Rochefort, he wrote to Thouin immediately, 14 July 1797: "I have succeeded in carrying the valuable breadfruit tree from the South Seas to Mauritius and Réunion. I have just arrived, coming from Mauritius, with 3 of those trees preserved from the danger of the sea."[287] He had to borrow money to pay for transporting his cases to Paris and found himself virtually unnoticed when he arrived. Given the turmoil of political events in Paris, the expedition had been long forgotten. André Thouin was not there to greet him, but was part of a team of scientists and artists summoned by Bonaparte to that part of Italy he had conquered. Poor Delahaye had to wait for 8 months before seeing Thouin again.

His reward, in 1798, was appointment as chief-gardener at the Trianon. Then, in 1805, he became gardener to the Empress Joséphine at La Malmaison, presumably recommended by Pierre-Etienne Ventenat, the director and curator of her gardens and brother of the deceased Louis. At Malmaison, Delahaye would cultivate some of the plants that had come from the expedition, and he became responsible for the introduction of several Australian and Tasmanian species of *Eucalyptus*, *Callistemon*, and *Acacia* into western Europe.[288]

Labillardière and the painter Piron did not leave Java until 29 March 1795 with the final shipment of expedition personnel: 50 men who benefited from another prisoner exchange. They reached Mauritius on 7 May 1795. As there was no immediate opportunity to proceed to France, Labillardière passed the time by exploring the island's higher mountains to observe their natural productions. There remains no further evidence about Piron's later life or death, an end as obscure as his beginning. He did entrust Labillardière with copies of his drawings made in the course of the expedition that would be used by Labillardière in the publication of his *Relation*. Several collections of his work may be found in the

286. Horner. 1995. *Looking for La Pérouse*, pp. 223-231.

287. Letouzey. 1989. *Le Jardin des Plantes*, p. 243.

288. Chevalier. 1953a. "Félix Delahaye." *Revue int. bot. appl.* 32: 63-64; Stafleu. 1967. *Adanson, Labillardière, De Candolle*, pp. 42-43.

Musée de l'Homme and in the Archives Nationales, but the number surviving must be greatly lower than his actual production on the voyage.[289]

The first ship available for the voyage to France, the *Minerve*, sailed on 21 November 1795. The governor gave her command to Patrice-Gaspard Laignel, one of Labillardière's companions in misfortune. They anchored at the Ile de Batz (Brittany) on 12 March 1796, from where Labillardière went directly to Paris. The worst possible news awaited him. The ship in the Dutch convoy containing the natural history collections had been captured by the British near St. Helena. The material had since been landed in England and was in enemy hands. As Labillardière had already come to believe that the botanical material was his, as the collector, he resolved to take all necessary steps to retrieve it.[290]

289. Collins. 1989. "Voir avec d'Entrecasteaux: l'oeuvre du dessinateur Piron," p. 57.
290. Labillardière. 1800. *Relation*. 2: 330-332.

XV
From Explorer to Closet Botanist

The first scholar to publish several of the letters relating to Labillardière's personal campaign to recover his collections advised his readers that Labillardière, in both his letters and in his book, had made charges against both Auribeau and Rossel that were "as serious as they were without foundation."[291] As that judgment was rendered without any supportive evidence, his readers might well have assumed that there had been no legitimate alternative to the royalist claim, as represented by Rossel in 1796, namely, that the collections belonged to the late king's legitimate heir, Louis XVIII.

In fact, the question of legal ownership was more complicated and unclear. Louis XVI had indeed sponsored the expedition, and it had been understood that all collections would be delivered to the crown. But, after the execution of Louis XVI and the overthrow of the monarchy, all crown properties had been nationalized. That action gave the French Republic, governed by the Directory in 1796, a legal claim to the collections. One of the directors, moreover, Louis-Marie Larevellière-Lépeaux, took a personal interest in the dispute as he was a botanophile and a close friend of André Thouin. The resolution of these conflicting claims was hampered by the intractable fact that republican France, at war with Britain, no longer recognized the legitimacy of the French monarchy; while the British recognized, not the French Republic, but the royalist regime in exile.

The validity of Labillardière's claim to the collections was more practical than legal. From the moment he learned on Java of the fall of the French monarchy, he had believed himself relieved of the obligation to relinquish his journal to Auribeau, and that the plant collections had become his as the collector. Accordingly, he regarded all subsequent attempts by royalist officers to acquire his property, not only as unjust, but as political reprisal. He could not then have known that such properties had been nationalized.

A few weeks after Labillardière and his fellow republicans had been arrested and confined in Samarang by the Dutch officials in collusion with Auribeau, Labillardière wrote a letter to Sir Joseph Banks about the ample collections he had made. He described his plight as having been the work of Auribeau and indicated that Auribeau had possession of his collections. He enclosed a similar

291. Edmond Bonnet. 1891. "Les Collections de l'expédition envoyée à la recherche de La Pérouse d'après documents inédits." *Association Française Avancement Science Centre Recherche* (Marseilles) 2: 489.

letter to L'Héritier de Brutelle in Paris, asking Banks either to forward it or send it to Desfontaines or Thouin to try to forestall calumny.[292] Given the circumstances, it is hardly surprising that the letter did not reach Banks until 20 months later, 4 February 1796, only five weeks before Labillardière's landing in Brittany.

Lt. Rossel, meanwhile, had been taken from another Dutch ship in the convoy and had been brought to London on 1 November 1795. He had protested at once against the British seizure of the collections on the reasonable grounds that he had been unaware that war had been declared between Holland and Britain and had assumed he was traveling on a friendly ship. The validity of the claim was admitted in time, and the 92 cases from the expedition that had reached England were put at the disposal of the legitimate king, Louis XVIII, then living in exile in Courland. The Admiralty, meanwhile, had taken advantage of the opportunity to examine and copy Beautemps-Beaupré's maps.[293]

Rossel's success in securing the French monarchy's claim on the collections, thus presumably guaranteeing their ultimate return to France, was short-lived. Louis XVIII instructed his ambassador in London, François-Henri, duc d'Harcourt, to offer the plant collections to Queen Charlotte. Harcourt wrote to Banks on 29 March 1796 to say that the queen had suggested that Banks review what was in the cases.[294]

In Paris, Labillardière began to make his own claim to the plant collections in a letter to André Thouin, 4 April 1796. Using angry, exaggerated language, he related the manner in which Auribeau had gone about putting an embargo on the two frigates, the collections, and all other materiel on board. He characterized Auribeau's motives in having the seven republicans removed from Surabaya to detention in Samarang, as having been a precaution to prevent true patriots from thwarting his schemes.

He also described a visit to Fort d'Anké by Beautemps-Beaupré and Renard on 19 November 1794, several weeks before the Dutch convoy sailed from Batavia. They had told him that Lt. Rossel wanted to reduce the bulk of the collections by taking only one or two specimens of each species in both Riche's and Labillardière's collections. Rossel had, therefore, given Deschamps the duty to remove the duplicates, which he was free to retain as his own property. While it seems probable that Rossel had indeed given such instruction, justifying Labillardière's outrage at the time he wrote the letter, to what extent the indolent Deschamps carried out

292. Labillardière to Banks, 6 April 1794. W. R. Dawson. 1958. *The Banks Letters*. London: British Museum, p. 514.

293. Sir Gavin De Beer. 1960. *The Sciences Were Never at War*. London: Nelson, pp. 45-48; Horner. 1995. *Looking for La Pérouse*, pp. 235-236.

294. Chevalier. 1953b. "Un grand voyageur normand," p. 198.

the instructions is unclear. Certainly, Labillardière would ultimately have duplicate specimens to distribute to colleagues in later years: to Banks, Jussieu, Robert Brown, and James Edward Smith in particular.

This letter to Thouin contained a statement that has usually been overlooked in assessing the dispute over ownership of the collections. Labillardière emphasized that he had taken care to collect a substantial number of duplicate specimens, "much more numerous than he had to deliver to the chief of the expedition," as he had expected to share examples of his plants with others.[295] Such duplicates he considered to be his private property, and Rossel had meant to divert them to the royalist Deschamps.

As the monarchy had fallen, Labillardière had concluded that he was no longer under any obligation to the crown. In any case, when writing to Thouin, he had no knowledge of how many of his specimens the cases still contained, nor whether they had been turned over to Rossel by the English. In view of the state of war between England and the Republic, he concluded, "I believe that to get them to return these materials to the Republic they must be made to realize that this is personal property."[296]

Accordingly, Labillardière wrote to Sir Joseph Banks a few days later. He accused the Dutch on Java as having authorized Rossel's theft of his collections. Aware that these collections then rested in British Customs, he asked Banks to do whatever he could to help him recover *his rightful property.* He implied, moreover, that if the collections were really to become a useful benefit to science, they must not pass into other hands. "Your love for the sciences is the surest guarantee I have that you will devote yourself to obliging me in the circumstance." He gave his address as in care of Desfontaines at the Jardin des plantes.[297]

Bank's response indicates that he had consequently reversed himself about the propriety of restoring the collections to Louis XVIII and promised his assistance in getting his government to correct the error. "Rest assured of my unwearied Diligence. That the science of two Nations may be at Peace while their politics are at war is an axiom we have learned from your Protection to Capt. Cook & surely nothing is so likely to abate the unjustifiable Rancour that Politicians frequently entertain against each other as to see Harmony and good will prevail among their Brethren who cultivate science."[298]

295. Labillardière to Thouin, 4 April 1796. Ibid. pp. 194-195.

296. Ibid. p. 196.

297. Labillardière to Banks, 14 April 1796. Ibid. pp. 196-197; De Beer. 1960. *The Sciences Were Never at War,* pp. 53-54.

298. Banks to Labillardière, 9 June 1796. De Beer. 1960. *The Sciences Were Never at War,* p. 55; Dawson. 1958. *The Banks Letters,* p. 906.

Good as his word, Banks at once sought an interview with Lord Grenville at the foreign office to persuade him that the collections were not the property of the French crown. The matter was awkward for Banks as he had previously recommended that Queen Charlotte be given one specimen of each plant species in the collections. She had accepted the offer on the assumption that the specimens were being offered to her by the King of France. It would now be necessary to explain to the queen why she must be deprived of what she had accepted.[299]

Banks left the interview under the impression that Grenville was inclined to consider favorably the return of the Labillardière collections and so informed the French official delegate in London.[300] When no such decision was forthcoming, Banks pressed the matter by warning Lord Grenville that a refusal to return Labillardière's collection would damage England's reputation after the generous treatment of Captain Cook by France.[301]

Three weeks later, Grenville anthorized Banks to return the collections to France, and Banks at once notified the French delegate in London that the collections could be shipped to the Jardin des plantes in Paris.[302] On the same date, he sent the same news to Jussieu, the director of the Muséum de l'Histoire Naturelle, with the assurance that the cases had not been opened. "I would not even look at the contents for fear of carrying away even a single botanical idea that belonged to a man who had gone out in their conquest at peril to his life." While he was anxious to learn about Labillardière's discoveries, he had not retained a single leaf or flower.[303]

The collections arrived at the Jardin des plantes in due course, shipped by Jean Charretié, who had properly thanked Banks on behalf of the Directory.[304] Jussieu held the cases unopened until Labillardière could return from Italy. Like André Thouin, he was serving on Bonaparte's commission that sought items of artistic and scientific interest. Thus, his own response to Banks was tardy. In view of his own pronounced views on his own property rights, it is notable that he made no moral appraisal of the commission's work in Italy: "I owe both to your friendship for me and to your love of justice everything you have done so that the fruit of the most dangerous possible voyage has not fallen into Rossel's possession."[305]

299. Dawson. 1958. *The Banks Letters*, p. 686.

300. Banks to Jean Charretié, 16 June 1796. Ibid. p. 212.

301. Banks to William Wyndham, baron Grenville, 20 July 1796. Ibid. p. 903.

302. Banks to Jean Charretié, 10 August 1796. Ibid. p. 212.

303. Banks to Antoine-Laurent de Jussieu, 10 August 1796. Ibid. p. 482; Chevalier. 1953b. "Un grand voyageur normand," p. 198; Antoine Lasègne. 1845. *Musée botanique de Monsieur Benjamin Delessert*. Paris: Fortin, Masson, p. 78.

304. De Beer. 1960. *The Sciences Were Never at War*, p. 64.

305. Labillardière to Banks, 8 December 1796. Chevalier. 1953b. "Un grand voyageur normand," p. 197.

Labillardière had not hesitated to record his views of Rossel's character in letters and would pursue that liberty when publishing his journal in 1800. For whatever reason, Rossel avoided such personal issues when editing the Entrecasteaux documents for publication in 1808, leaving his side of the story dependent upon his actions. Two major watersheds divided them, one political, the other intellectual. Labillardière was republican and a naturalist, Rossel a royalist and a navigator.

Rossel's efforts to preserve the collections and records of the expedition for the crown did not reveal where his own intellectual interests lay. That evidence had begun to emerge at the Cape of Good Hope when Rossel had assumed the astronomer's duties after the departure of Jean Blavier in 1792. He would handle those duties very competently. That interest and competence became manifest only when Rossel edited the records of the expedition.

Rossel had remained in England for about seven years, momentarily as a prisoner, soon after as an émigré linked to the duc d'Harcourt. The navigational charts he had possessed were returned to him after the British Navy had examined and copied them. In the exchange of prisoners arranged after the Treaty of Amiens in 1802, Rossel was free to return to France and did so in 1803. He took with him the journal of Rear Admiral d'Entrecasteaux; the maps and drawings; and the written observations of the hydrographer Beautemps-Beaupré; of François Achard de Bonvouloir, a naval-officer candidate and student of astronomy who had died on Java; of Ensign Jacques-Bertrand Legrand; and of Ensign J.-B.-Philibert Willaumez. These materials served him well in preparing the two volumes for the official publication in 1808 of the Entrecasteaux expedition. The volumes generated little public interest thanks to the prior publication of Labillardière's account. The latter had enjoyed widespread readership after 1800, had appeared almost simultaneously in an English translation, and had evidently captured the market.[306]

The second volume of Rossel's work was divided into two parts. The first section was a survey of the astronomical observations made during the voyage, including the methods of correction that can be used either on land or at sea in order to establish geographical positions. The second part encompassed the data and the outcome of those same astronomical observations, thus assembling most of the facts upon which rested the discussion of principles provided in the first part.[307]

306. Ibid. pp. 191, 199.
307. Rossel. 1808. *Voyage d'Entrecasteaux*. 2: i.

The devotion of an entire volume to reflecting the astronomical and navigational record of the expedition, while giving no space whatsoever to the achievements in the natural sciences during the voyage, testified indisputably, not only to Rossel's particular competence, but to what he believed the expedition's chief purpose and importance had been. From such a point of view, one might readily have regarded the bulky collections of natural history specimens on board ship as excess baggage, to use the words in Ensign Jurien's *Souvenirs*.

Even so, Rossel had done his duty by escorting those collections back to Europe. He may even have been dismayed when, after all his efforts, Louis XVIII saw fit to offer them to Queen Charlotte, a matter on which he remained silent. Yet, he can hardly have appreciated the scientific value of those collections. In his own realm, however, he was a scientist and recognized as such by election to the Académie des sciences in 1812.

Both Rossel and Labillardière published vocabularies of native languages in appendices to their books. While the two sets of vocabularies are not identical in every instance, suggesting that more than one person had taken an interest in recording native words, there is also sufficient duplication to indicate either collaboration or wholesale borrowing.[308]

Labillardière's journal featured careful and valuable observations of native customs, foods, and physiques, made with a clinical precision that derived from the botanist's capacity for discrimination. There was never a hint of condescension; and whether referring to them as *Sauvages* or *Naturels*, he always used a capital letter. A stopover was frequently too short to make all the observations he wanted, especially considering the intimate details that had been requested by the Société de Médecine, the exceptions being on Tasmania and New Caledonia. Sojourns on the Dutch islands of Ambon and Java had been lengthy, to be sure, but there the indigenous populations had been under Dutch and Chinese influences for well over a century. No naturalist could presume that a pristine people or culture had been preserved.[309]

Labillardière's *Relation du voyage à la recherche de La Pérouse* revealed his pronounced practical approach to Nature. Whereas his other publications would be devoted to the description of new genera and species, the *Relation* contained many passages on vegetable food-stocks, not merely the breadfruit tree. Moreover, even though he knew of the development of a natural classification system, he preferred to use the more directly useful sexual system, making him a true Linnaean. A more philosophical spirit would surely have made him a Jussiaean. While he had

308. Ibid. 1: 552-590; Labillardière. 1800. *Relation*. 2: 3-61.
309. Chevalier. 1953b. "Un grand voyageur normand," p. 119.

certainly been affected by his century's teaching on the benefits of natural life, he was not blinded to its actualities: the slavery and cannibalism born of continual warfare. Nevertheless, his descriptions did not ring of a sense of superiority or reveal any missionary spirit. He might deplore the inequality of the dangerous hardships endured by the Tasmanian women, but his remedy did not go beyond urging the men to share them.[310]

By publishing his *Relation* in 1800, Labillardière could not use the nautical material held by Rossel in London. He did have Ensign Legrand's nautical journal, and Piron had given him duplicates of drawings before leaving Java. A few of the illustrations of plants were signed by Pierre-Joseph Redouté, the preeminent plant painter of that day.

Those who had charged, during the voyage, that Labillardière was planning to publish a radical republican account of the expedition to refute an official account were proved to be wrong. Frank Horner has made the valid point that Labillardière's book turned out to be more objective than Rossel's work. Labillardière did, indeed, argue the naturalists' case where he believed circumstances required it; he did not conceal some of the navigational lapses as Rossel would; but there was nary a hint of disloyalty to either Entrecasteaux or Huot.[311]

The publication of the *Relation* brought Labillardière election to the Académie des sciences on 26 November 1800 to succeed the recently-murdered L'Héritier de Brutelle, one of Labillardière's few close friends. After his election, Labillardière would become one of the most dedicated members of the Académie, a position that provided him an income for life.[312] The election was all the more a triumph as it was won despite the vigorous opposition of Etienne-Pierre Ventenat, a man of the most modest botanical accomplishments, but who had a pronounced talent for linking his name with the distinguished work of others. He denigrated routinely the accomplishments of those whose scientific experience quite surpassed his own. His botanical career, incidentally, had begun with his defection from a religious order in 1789 and never encompassed any fieldwork.

Labillardière must have turned rather promptly to preparing his new species from Western Australia and Tasmania for publication as they appeared between 1804 and 1806.[313] The two volumes included the descriptions of 265 new species, accompanied by 265 plates from paintings by Piron, Pierre-Joseph Redouté, and one of his pupils, Pierre-Antoine Poiteau. Some additional species, of course, had

310. Stafleu. 1967. *Adanson, Labillardière, De Candolle*, pp. 35, 41.

311. Horner. 1995. *Looking for La Pérouse*, pp. 244-245.

312. Auguste de Saint-Hilaire. 1834. "Discours prononcé le 10 janvier 1834 sur la tombe de M. de La Billardière." *Annales des sciences naturelles. Botanique* ser. 2. 1: 41.

313. Labillardière. 1804-1806. *Novae Hollandiae Plantarum Specimen*, 2 vols. Paris: Dominae Huzard.

been published previously in his *Relation*.[314] Some of the locations Labillardière cited for his species have had to be corrected in recent years. At best, they were very general; and, by 1804, he had seen other collections that probably acted to confuse his memory, as considerable time had passed since his own collections had been made. Some of his mislocations within Tasmania had been recognized by Joseph Dalton Hooker as early as 1860.[315] A recent attempt has also been made to identify occasional Tasmanian plants described in Labillardière's *Relation*, but never named or published by him, identifications acknowledged to be only probable. *Caveat lector*.[316]

There would be a lapse of twenty years between the publication of his Australian plants and his New Caledonian plants, all the more surprizing as Labillardière had been the first European to collect on that island; and one might have expected some haste in establishing that priority. The initial diversion was some unfinished business. The publication of his *Icones Plantarum Syriae Rariorum* in five decades, begun in 1791, had been interrupted by the Entrecasteaux voyage. Labillardière returned to it after 1806 and completed publication in 1812 with Pierre-Antoine Poiteau as illustrator. Even so, that it required a dozen more years, uninterrupted by either teaching obligations or by the responsibilities of a family, suggests that a peculiar inertia had gripped him. When the work finally appeared in 1824-1825, it featured 80 new species, all illustrated by Pierre-Jean-François Turpin, another pupil of Redouté.[317]

While Labillardière has been given all due credit for the publication of the *Relation* and three major floras, it has since been recognized that he left much of his material untouched despite the 40 years he lived in Paris following the expedition. His personal herbarium amounted to more than 4,000 plants; he described only about 400. That may partially account for why he was not as much appreciated in scientific circles during his lifetime as he merited. He was ranked below his friend Desfontaines, who held a professorship at the Jardin des plantes; and certainly below Antoine-Laurent de Jussieu whose system of natural classification had set him above and beyond all others. Indeed, Labillardière's failure to rally to natural classification in his work merely advertized his relative indifference to

314. Sir Joseph Dalton Hooker. 1860. *The Botany of the Antarctic Voyage of H.M. Discovery Ships Erebus and Terror, in the Years 1839-1843*. Part III, *Flora Tasmaniae*, 2 vols. London: Lovell Reeve, p. cxiii.

315. E. Charles Nelson. 1974. "The Locations of collection and collectors of specimens described by Labillardière in *Nova Hollandiae Plantarum Specimen*: additional notes." Papers & Proceedings of the *Royal Society of Tasmania*. 108: 162, 167-168.

316. Brian Plomley & Josiane Picard-Bernier. 1993. *The General: the visits of the expedition led by Bruny d'entrecasteaux to Tasmanian Waters in 1792 and 1793*. Launceston: Queen Victoria Museum, pp. 233-239.

317. Labillardière. 1824-1825. *Sertum austro-caledonicum*, 2 parts. Paris: Dominae Huzard.

botanical theory. In the heyday of Jussieu and Candolle, such negligence reduced Labillardière to lower rank. His work represented the classic preoccupation of the later 18th century with the discovery and inventory of new species, not the newer systematics of the early 19th century. It is the more regrettable that he never treated the plants he collected on various archipelagos that have remained rarely visited into the 20th century.[318]

To emphasize what he failed to accomplish would be to risk devaluing unfairly what he did accomplish. His floras were major contributions, and his care in forming new generic and specific names was the mark of a well-trained and discriminating mind. Most of his plant names were formed from Greek words and generally based upon well-chosen characters. He frequently demonstrated his technique in the *Relation*. *Dracophyllum*, for instance, referred to a calyx composed of six small, oval, scale-like leaves, such as the scales of a serpent. He gave the type species the name *Dracophyllum verticillatum* Labill., because its flowers were arranged in a ring.[319] He on occasion formed an imaginary Greek name through ellipsis. *Campylonema*, for example, referring to bent or curved filaments of the stamens, was ellipsized into *Campynema* Labill., a new genus in the Amaryllidaceae.[320] W. T. Stearn recognized that similar ellipsis would be used by Robert Brown in his *Prodromus Florae Novae Hollandiae et Insulae Van Diemen* in 1810.[321]

Because the significance of plant distribution had not yet been generally recognized in the early 19th century, the fact that Labillardière had shown himself to be a pioneer in phytogeography in the *Relation* was evidently either overlooked or undervalued in the Paris of his day, only recognized by Thomas Hart in 1954. Moreover, his interest in native food plants and native techniques in their cultivation was an additional field that would only be recognized as an important contribution in a later day.

But how are we to account for the peculiar inertia that seemed to grip Labillardière during those nearly 40 years he lived in Paris following the expedition? The friendship he had formed with René Desfontaines about 1780 lasted for life. Despite the differences in their careers, less easily perceived similarities probably account for the long attachment. Both went on long research trips overseas, and both became, in later years, notably reclusive. Once Desfontaines had been

318. Chevalier. 1953b. "Un grand voyageur naturaliste normand," pp. 108-109; Pierre-Jean-Marie Flourens. 1837. "Eloge historique de Jacques-Julien de Labillardière." *Annales du Muséum d'histoire naturelle*. 16: xxvi-xxxi.
319. Labillardière. 1800. *Relation*. 2: 211. Tab. 40.
320. Labillardière. 1804. *Nov. Holl. Pl.* 1: 93. tab. 121.
321. Thomas S. Hart. 1954. "Labillardière's plant names." *Victorian Naturalist*. 70, January: 173-175; Stafleu. 1967. *Adanson, Labillardière, De Candolle*, p. 37.

appointed to a professorship of botany, he settled into the Jardin des plantes from which he rarely emerged; and once Labillardière returned from the voyage with Entrecasteaux, he became unsociable and reclusive. To be more precise, both men restricted their sociability to fellow naturalists.

At least, that was the impression given. Chevalier mentioned, with some hesitation, that Labillardière had been characterized after his death as a great misanthrope by a medical correspondent in *le Progrès médical* of January-April 1841: "He chose to live on the sixth floor so as not to have to receive visits, and during the fine season he lived in a sort of hut on the outskirts of Paris. Only his friend René Desfontaines could come to visit him. . . . He always missed his friends the aborigines."[322]

Chevalier knew that the observation contained a grain of truth, but recognized it also to be the superficial obervation of a critic with little insight into Labillardièr's character. Certainly Labillardière had often felt a sympathy for aboriginal peoples, but he never evinced the least aspiration to join them permanently. Only an enemy of one's own civilization would commit such an act, and Labillardière was far too commited to the sciences of his own European civilization to have entertained any idea of flight. We need only to recall his disapproval of the unhappy Tongan, who sought to sail away with the French, to realize his recognition of the impossibility of bridging such a profound cultural chasm successfully.

The mystery of Labillardière's inertia after an energetic long voyage cannot be satisfactorily resolved for lack of abundant evidence about his character. Aside from his scientific publications, he wrote very little. A search of public archives years ago for his letters produced very few, and they were published as a contribution to understanding his psychology. The very slimness of his correspondence is, of course, a suggestion that he could always have been a man of considerable reserve, or that he limited his friendships to people of professional interest.

A letter dated 13 March 1788, written during his Near Eastern trip, to an unknown correspondent in Montpellier, revealed a lively interest in earlier friends from medical school days: "Do you hear anything about our friend M. Gribert? Please give news of me to M. de Lermina, to M. Bernard, etc. Has there been another trip interesting for natural history since my departure, whether in France or abroad? Also tell M. Broussonnet about me. If the streams of Crete yield some fish I do not know, I will make him a present of them. Perhaps they will not be new to him."[323] A letter he sent to Banks with a copy of his *Relation* was not only

322. Chevalier. 1953b. "Un grand voyageur naturaliste normand," p. 105.
323. Ibid., pp. 190-193.

a gracious acknowledgment of Banks' efforts to obtain the return of the captured collections, but the letter of a man dutiful to social obligations.[324]

Various bits and pieces of evidence amount to consistent testimony to a scrupulous honesty, to a notable freedom from selfish motives, to that disinterestedness we always hope to find in a life dedicated to scholarship. In 1809, for instance, he received a grant from the Académie des sciences to enable him to have his Syrian plants drawn; and between 1808 and 1822 additional modest grants to cover expenses for the copperplate engravings for his illustrations. His dossier at the Académie still contains his precise accounting for all expenses incurred.

After the Entrecasteaux expedition sailed from Brest in the autumn of 1791, Labillardière had made no attempt to conceal his republicanism. Yet, he was no fanatical partisan, hostile to all royalists, as his friendly references to the young naval officer-candidate and budding astronomer, Achard de Bonvouloir, repeatedly testified. If he bore a lasting grudge against Rossel, the issue had not only been a dispute over the ownership of the collections, but because Rossel had had him imprisoned for his political opinions. As for Rossel, he did not forgive Labillardière for having denounced him in the *Relation*; and they cannot have been congenial colleagues in the Académie after Rossel's election in 1812.[325]

At Labillardière's funeral in 1834, Saint-Hilaire not only recalled that Labillardière had been one of the most dedicated members of the Académie, but that his voyages, long and difficult, had proved his ardor and courage; and that he had always been distinguished by his love of justice and his moderation. Moreover, that he had been a witty and good-hearted man, generous with his money. The basis of his character had been a noble independence.[326]

The formal *éloge* read at the Académie by Pierre Flourens in 1835 seems to have confirmed the above characterization: "I have known M. Labillardière personally. I pay tribute to all the virtues with which he was endowed: to his upright and elevated soul, to his concern for the public well-being, to his friendship that was firm and devoted. Labillardière never changed; he was always Labillardière: a man of probity, a wise and just man with an unpretentious character. . . . He was totally dedicated to justice, and independence was his dominant character trait. He lived alone and was dependent upon no one. He could be a genuine friend, but was discriminating or circumspect in his friendships, always quick to take offense at the slightest suggestion of intrusion."[327] Discriminating and circumspect in his friendships implies a disinclination to socialize beyond the limits

324. Labillardière to Joseph Banks, 5 March 1800. Dawson. 1958. *The Banks Letters*, p. 514.
325. Chevalier. 1953b. "Un grand voyageur naturaliste normand," pp. 105-107.
326. Saint-Hilaire. 1834. "Discours prononcé le 10 janvier 1834," pp. 41-42.
327. Flourens. 1873. "Eloge historique," p. xxx.

of his intellectual realm. Reclusive is perhaps the word, but the charge of "great misanthrope" seems unfounded.

The matter comes down, therefore, to the phenomenon of the botanist-voyager who undergoes years of hardship in the face of constant danger, showing great talent for collecting and recording under demanding circumstances, while developing such a high degree of self-reliance and independence that he becomes essentially unfit for life in settlements. Paris was no longer his congenial home, and his productive energies suffered accordingly.

The number of years he required to bring forth three works, and the amount of plant material left untouched, suggests the onset of an emotional paralysis against which he had had to fight in order to bring any project to fruition. That phenomenon has not been experienced by every botanist-voyager, Joseph Dalton Hooker for example; but the pattern is a compelling one and frequently observed. Perhaps that is what Labillardière meant when he remarked that he missed his friends the aborigines: he would rather have been traveling.

Jussieu's readiness to turn over Labillardière's collections to him, rather than simply making them easily available to him at the Muséum d'Histoire Naturelle, ultimately had a regrettable outcome. Some partial sets of duplicates were given to interested botanists by Labillardière, but he kept his main collection for his lifetime. After his death, his personal herbarium was acquired by Philip Barker Webb, an Englishman who lived in Paris during his later years and was an amateur of Mediterranean civilization. Webb bequeathed his collections to the Grand-Duke of Tuscany. Since about 1854, as a consequence, the Labillardière collection has constituted the Webb Herbarium at the Instituto Botanico in Florence.[328]

Because of the earlier distribution of some duplicates, Labillardière's specimens did turn up in other herbaria. About 300 species from Australia-Tasmania can be found in the Candolle Herbarium in Geneva. The Delessert Herbarium, also in Geneva, contained not only Labillardière's plants from Syria, but diverse specimens acquired through the purchase of the herbaria of Lambert and Dr. Lemonnier. There are also diverse specimens in the Muséum d'Histoire Naturelle in Paris,[329] where, by all rights, the full collections ought to be: the dessicated legacy of a dedicated life in natural science.

328. Stafleu. 1967. *Adanson, Labillardière, De Candolle*, p. 41. Webb was a coauthor of *Phytographia canariensis* published in Paris, 1836-1850.

329. Lasègne. 1845. *Musée botanique de Delessert*, pp. 56-57, 75-78; Alphonse de Candolle. 1880. *La Phytographie*. Paris: G. Masson, p. 425.

XVI
Epilogue

No trace of La Pérouse had been found by the Entrecasteaux expedition. In May of 1793, when Entrecasteaux passed from New Caledonia into the Santa Cruz group, a bearing had been taken on an unknown island in the distance that he named the Ile de la Recherche, now Vanikola Island. We cannot say precisely how close they came to the site that would prove to be the locale of the La Pérouse shipwreck, nor is it certain why they left the island uninvestigated.

Frank Horner offered the reasonable conjecture that the loss of time Entrecasteaux suffered early in 1792, when he chose to investigate the rumor about the Admiralty Islands, only to have to backtrack to his prescribed route, led necessarily to later shortcuts. In particular, he cited a pertinent paragraph in Fleurieu's instructions that recommended, if possible, that Entrecasteaux, after leaving New Caledonia, should proceed to the archipelago of the Espirita Santo (now the New Hebrides) in order to investigate several small islands to the northeast of that archipelago: islands that had been discovered by Captain Bligh, but which had escaped the attention of both Bougainville and Cook. It is again reasonable to conjecture, as Horner did, that had Entrecasteaux made that recommended exploration he would have subsequently come close to Vanikola Island. On neither of his passages through those waters did he make the recommended exploration, whether for lack of time or because of contrary winds; and the opportunity to fulfill the primary mission of the expedition was missed.[330]

Years went by without any news of La Pérouse until, in 1827, the British East India Company. sponsored a new search by Captain Peter Dillon. He would publish an account of his voyage, a marvelous story wonderfully told, from which one can tell that he had prepared for his adventure by studying the narrative by La Pérouse edited by Milet-Mureau, as well as La Billardière's account of the Entrecasteaux expedition. Although he did not mention Beautemps-Beaupré by name, he acknowledged the use of charts that had to have been based upon his; and we know that the British had copied them. In sum, Dillon knew La Pérouse's

330. Rossel. 1808. *Voyage d'Entrecasteaux*. 1: xxxi-xxxii; Horner. 1995. *Looking for La Pérouse*, pp. 256-257.

projected route after leaving Botany Bay, and, therefore, proceeded to the Tonga Islands to begin his search, then sailing on northwestward.[331]

He reached Rotuma on 1 September 1827, an isolated island north of the main Fiji chain, then sailed westward for Tikopia Island in the Santa Cruz group. During the nearly 20 years Dillon had been at sea, he had visited Tikopia at least once before. On this second occasion, he had on board Martin Bushart, one of those renegade Europeans increasingly found on Pacific islands in the 19th century, who Dillon had picked up on Tongatapu Island, and who had a Tikopian wife. Together, they might serve as interpreters.

Arriving off Tikopia on 5 September 1827, Dillon put Bushart ashore to open some communication with the natives. Bushart learned that there were 5 Englishmen on the island, probably escapees from New South Wales; but he had had the good luck to meet Rathea, a native who had returned from Vanikola about three or four months earlier, having resided there for about 5 years. He had brought back chain-plates, long iron bolts, and a crowbar.

From Rathea, Dillon was able to glean the critical information that two ships had been wrecked during the night on reefs considerably off-shore from Vanikola, but long ago when Rathea would have been only a child of 8 or 10. He related that one of the ships, a total wreck, had subsequently drifted to shore near the village of Vanou. The second ship had been driven ashore near the village of Peou, but somewhat more intact. Sometime much later, Rathea had seen large pieces of iron at the sites, too heavy to have been moved.

Captain Dillon, having recognized from his maps that the Entrecasteaux expedition had passed Vanikola Island, perhaps within 27 to 30 miles, but without reconnoitering it, at once suspected that Entrecasteaux had unwittingly bypassed the probable locale of the La Pérouse disaster. Consequently, he had Bushart persuade the head chief on Tikopia to order his people to bring in all ironwork in their possession, or anything else that had come from Vanikola, for possible purchase by Dillon. Rathea, then estimated to be about 50, volunteered to go with Dillon to Vanikola as pilot and interpreter.

They reached Vanikola on 7 September 1827, finding the island well-guarded by dangerous reefs. During a lengthy effort to find a safe anchorage, initial communication with shore was made by small boats. Thanks to Rathea's presence, the natives were easily convinced the visit was peaceful. Nonetheless, persistent attempts to interview inhabitants were extremely difficult and made worse by the

331. Peter Dillon. 1829. *Narrative and successful result of a voyage in the South Seas performed by order of the government of British India, to ascertain the actual fate of La Pérouse's expedition*, 2 vols. London: Hurst, Chance & Co. 1: viii, lx.

fact that Bushart, Prussian by birth, had a poor command of English, nor had he really mastered the native language.

It soon became apparent, however, that virtually every man on the island possessed an iron adze of native manufacture, but from material procured from the wrecked ships. Over the next few days, a sustantial number of objects of European origin were acquired, either through trade or purchase. One notable item was a silver vessel resembling a sauceboat with a fleur-de-lis stamped on it in two places.

From the mass of conflicting testimony, it was deduced that the ship, which had struck the reefs off Peou, had broken into pieces that drifted ashore. Bodies were found, many of them mutilated by sharks. Four survivors had remained temporarily with the natives, but had soon joined the more numerous survivors of the second ship wrecked off Vanou. The latter had built a substantial stockade for defense, within which they were building a small ship out of salvaged wreckage. This small ship was said to have departed five moons after the shipwreck: a two-masted ship carrying all survivors except two, one a white chief, the other his servant. If Rathea was correct, the little ship was never again seen by the Vanikola islanders.

On 19 September 1827, Dillon ordered an exploration of the island's coastline to look for additional evidence. This circumnavigation, carried out by small boats, produced many aged artifacts. Near Peou, they encountered a piece of ground, about an acre in size, adjacent to a small stream, that had obviously been cleared in the past: the best place on the island adapted to building and launching a ship. At the village of Amma, where they went ashore to replenish water-kegs, they saw a decaying wooden plank placed on its edge across the doorway of a nearby dwelling to keep out the pigs. A fleur-de-lis and other ornamental work could still be seen on it. It could have been part of a stern that, when intact, exhibited the royal arms of France. Dillon was able to trade a hatchet for the plank. The discovery of a European millstone was believed to be consistent with the provisions on the La Pérouse expedition, which had carried kiln-dried grain.

Having collected enough evidence to prove that Vanikola had been the island where La Pérouse had perished, and aware that no trace of the survivors' little ship had ever been found, Dillon made preparations to leave on 8 October 1827. He had earlier promised both Rathea and Bushart that he would return them to Tikopia Island; but the ship's surgeon prevailed upon him, in view of a large number of sick crewmen, to sail as directly as possible to New Zealand for urgent treatment. This violation of his parole clearly distressed Dillon who, soon after reaching New Zealand on 5 November 1827, began making arrangements with a Captain Kent of the brig *Governor McQuarie* to pay for the passage of his interpreters to Tongatapu and Tikopia. One of his own ship's officers, John Russell, agreed to make the trip to reassure the interpreters that the pledge would be kept. But poor Rathea, despairing that he would not see his homeland again or eat its familiar food, sickened and died on 7 December 1827, a death that seared Dillon's conscience. The others departed with Captain Kent four days later.

Once the refitting of his own ship was completed, Dillon sailed on 13 December 1827. He took his findings to the East India Co. in Calcutta. Then, in May of 1828, he departed with his collection for Europe where he had not been for 22 years.[332] The following February, the relics were examined in Paris by Barthélemy de Lesseps, the Russian interpreter who had left the La Pérouse expedition at Kamchatka, and found to be authentic. Dillon was also able to talk to Rossel, by then a rear-admiral and near the end of his life. An audience was granted by Charles X in March of 1829. Dillon was made a chevalier in the Légion d'Honneur, awarded 10,000 francs to compensate him for the voyage from India to France, and granted a annual pension of 4,000 francs for life.[333]

It has been said that Louis XVI, on the eve of his beheading in 1793, had asked whether there had been any news of La Pérouse. The story may be apocryphal, but it has been credible as a reliable reflection of that king's priorities. The answer to his question would not be forthcoming for another 35 years; and even then it would be incomplete without any knowledge of the fate of the small ship launched from Vanikola. Beyond which the king was denied any knowledge of the benefits and the costs of those expeditions he had sponsored in the hope that they would be the most glorious accomplishments of his reign. Even a man as limited as Louis XVI could share the aspirations of his enlightened age.

The costs in human life had been distressingly high, and the figures that remain are at best approximate. La Pérouse had left Brest in 1785 with about 239 souls on his two frigates. We know of 2 desertions during the sojourn in Chile, and of at least 6 other individuals who arranged to leave the expedition at a European outpost. The remaining 231 all perished, whether in the swamping of the longboats along the Alaskan coast, in the massacre on Samoa, or as a result of the shipwreck off Vanikola Island.

As for the Entrecasteaux expedition, we are dependent upon figures cited by Labillardière. The expedition had sailed in 1791 with about 219 people on board. By the time Labillardière reached Mauritius on his trip home, he knew of 89 fatalities. Very few of the deaths had occurred during the actual expedition;

332. Ibid. 2: 91-121, 139-146, 193-211, 241-243, 271-281, 306-355.
333. Horner. 1995. *Looking for La Pérouse*, p. 261.

most of them could be attributed to the long confinement on Java.[334] That figure does not include men like Claude Riche who, permanently debilitated by that experience, would die prematurely. In retrospect, one can sympathize with those who had preferred to cut and run toward Mauritius no matter the odds rather than accept Dutch hospitality on Java. By declaring the expedition to have ended with the arrival at Surabaya, thus bringing his published narrative to a close, Rossel did not undertake an accounting of the fatalities.

What had been the benefits from the two expeditions? The results had been unequal, of course, because of the loss of so much material in the La Pérouse shipwreck. Yet, when considered collectively, the expeditions had returned detailed discussions of astronomical observations made both upon land and upon sea. The maps and the elevations, especially those of Beautemps-Beaupré, including the charting of reefs and channels, were generally recognized as important contributions to safe navigation.

Observations made of native customs, languages, diseases, industries, agriculture, diets, and physiques were of immense interest to Europeans of that day and remain useful resources in our own, notably about groups whose customs have since been lost, or the groups themselves have disappeared. The naturalists, finally, brought home hundreds of unknown or little-known species that contributed greatly to the refinement of taxonomy, especially that of plants.

Did the benefits justify the costs? The answer seems to depend upon which generation asks the question. The postmodern answer might be that no new knowledge is worth the sacrifice of a human life. The 18th century could lament the loss of life (Labillardière called the fatalities on his expedition "saddening") while applauding the heroic dedication that extended its horizons. For that generation, the life without the continual search for new knowledge was a life not worth living.

334. Labillardière. 1800. *Relation*. 1: xv-xvi.

Bibliography

Adanson, Michel. 1757. *Histoire naturelle du Sénégal,* Avec la relation abrégée d'un voyage fait en ce pays, pendant les années 1749, 50, 51, 52 & 53. Paris: C. J. B. Bauche.

Aubert du Petit-Thouars, L.-M.-Aubert. 1804. *Histoire des végétaux recueillis sur les îles de France, La Réunion (Bourbon) et Madagascar.* Paris: Mme Huzard.

~ 1805. *Histoire des végétaux recueillis dans les îles australes d'Afrique.* Paris: Levrault, Schoell.

~ 1806. *Histoire des végétaux recueillis dans les îles australes d'Afrique.* Paris: Tourneisen fils.

~ 1806. *Genera Nova Madagascariensis,* secundum Methodum Jussiaeanum Disposita. Paris: n.p.

~ 1809. "Extrait de trois Mémoires lus à la première classe de l'Institut, sur l'histoire des plantes Orchidées des îles australes d'Afrique." *Nouveau Bulletin des Sciences,* par la Société Philomatique. Paris: April 1809. 1, no. 19: 314-319.

~ 1811. *Mélanges de botanique et voyages.* Paris: Arthus Bertrand.

~ 1822. *Histoire particulière des plantes orchidées recueillis sur les trois îles australes d'Afrique; de France, de Bourbon, et de Madagascar.* Paris: Arthus Bertrand.

~ 1825. *Notice historique sur la pépinière du Roi au Roule.* Paris: l'Imprimerie du Gueffier.

Audouard, Jean. 1910. *Un drame passionnel à la fin du XVIIIè siècle: le crime du marquis d'Entrecasteaux.* Paris: H. Daragon.

~ 1912. *Une famille provençale au XVIIIè siècle, les Bruny d'Entrecasteaux.* Paris: H. Daragon.

Bénézit, Emmanuel, ed. 1948-1955. *Dictionnaire critique et documentaire des peintres, sculpteurs, et graveurs,* 8 vols. Paris: Librairie Gründ.

Bentham, George. 1863-1878. *Flora Australiensis, a description of the plants of the Australian Territory,* 7 vols. London: Lovell Reeve & Co.

Bonnel, Ulane, ed. 1989. *Fleurieu et la marine de son temps.* Paris: Economica.

Bonnet, Edmond. 1891. "Les Collections de l'expédition envoyée à la recherche de Lapérouse d'après documents inédits." *Association française Avancement des Sciences Centre Recherche, Marseille* 2: 488-492.

Boreau, Alexandre. 1845. "Notes sur la vie et les travaux du botaniste Aubert du Petit-Thouars." *Bulletin de la Société industrielle d'Angers* 16: 55-66.

Bourgoin, Jean. 1989. "Fleurieu et le Voyage de Lapérouse." Ulane Bonnel, ed., *Fleurieu de la marine de son temps,* pp. 63-89. Paris: Economica.

Browne, Janet. 1995. *Charles Darwin Voyaging, A Biography*. Princeton: Princeton University Press.

Cap (Paul-Antoine Gratacap, called). 1854. *Le Muséum d'histoire naturelle*. Paris: L. Curmer.

Carr, S. G. M., and D. J., eds. 1981. *People and Plants in Australia*. Sydney: Academic Press.

Chevalier, Auguste. 1953a. "Félix Delahaye, le jardinier de l'expédition envoyée à la recherche de la Pérouse." *Revue internationale de botanique appliquée et d'agriculture tropicale* 32: 62-64.

～ 1953b. "Un grand voyageur naturaliste normand, J.-J. La Billardière (1755-1834)." *Revue internationale de botanique appliquée et d'agriculture tropicale* 32: 97-124, 185-202.

Collins, Roger. 1989. "Voir avec d'Entrecasteaux: l'oeuvre du dessinateur Piron." Ulane Bonnel, ed. *Fleurieu et la marine de son temps*, pp. 53-62. Paris: Economica.

Cordier, Henri. 1916. "Deux compagnons de Lapérouse [Lamartinière et Clonard]." *Bulletin de la Section de géographie (Comité des travaux historiques et scientifiques)* 31: 54-82.

Dakin, S. B. 1954. *The Perennial Adventure, a Tribute to Alice Eastwood, 1859-1953*. San Francisco: California Academy of Sciences.

Dawson, W. R. 1958. T*he Banks Letters*. London: British Museum.

De Beer, Sir Gavin R. 1960. *The Sciences Were Never at War*. London: Nelson.

Desfontaines, René Louiche. 1798-1799. *Flora atlantica, sive Historia Plantarum quae in Atlante, Agro tunetano et algeriensi crescunt*, 2 vols. Paris: L. G. Desgranges.

Dillon, Peter. 1829. *Narrative and successful result of a voyage in the South Seas performed by order of the government of British India, to ascertain the actual fate of La Pérouse's expedition*, 2 vols. London: Hurst, Chance & Co.

Dunmore, John. 1965. *French Explorers in the Pacific*, 2 vols. Oxford: Clarendon Press.

Dupont, Maurice. 1983. *D'Entrecasteaux: rien que la mer, un peu de gloire*. Paris: Editions Maritimes et d'Outre-Mer.

Dupree, A. Hunter. 1988. *Asa Gray, American Botanist, Friend of Darwin*. Baltimore & London: Johns Hopkins University Press.

Entrecasteaux. See Rossel 1808.

Flourens, Pierre-J.-M. 1837. "Eloge historique de Jacques-Julien de Labillardière." *Annales du Muséum national d'histoire naturelle* 16: xxvi-xxxi.

～ 1849. "Eloge historique d'Aubert Aubert du Petit-Thouars." *Mémoires de l'Académie des Sciences* (Eloges historiques) 20: 283-323.

Friis, Ib, and Finn N. Rasmussen. 1975. "The Two alternative systems of nomenclature proposed and used for orchids by Du Petit-Thouars, with special regard to *Corymborkia* Thouars." *Taxon* 24 (May): 307-318.

Frost, Alan. 1996. "The Antipodean Exchange: European Horticulture and Imperial Designs." Miller & Reill, eds. *Visions of Empire*. Cambridge: Cambridge University Press.

Gaziello, Catherine. 1984. *L'Expédition de Lapérouse 1785-1788, réplique française aux voyages de Cook*. Paris: Comité des Travaux Historiques et Scientifiques.

Gillispie Charles C. 1980. *Science and Polity in France at the End of the Old Regime*. Princeton: Princeton University Press.

Hart, Thomas S. 1954. "Labillardière's plant names." *Victorian Naturalist* 70 (January): 173-175.

Henley, Paul. 1999. "A Generous Quantity of Fat." *London Review of Books* 21, no. 17 (2 September): 24-26.

Hooker, Sir Joseph Dalton. 1860. *The Botany of the Antarctic Voyage of H. M. Discovery Ships Erebus and Terror, in the Years 1839-1843*. Part III. *Flora Tasmaniae*, 2 vols. London: Lovell Reeve.

Horgan, Paul. 1941. "Introduction." Josiah Gregg. *Diary & Letters*. Norman: University of Oklahoma Press.

Horner, Frank. 1995. *Looking for La Pérouse, d'Entrecasteaux in Australia and the South Pacific 1792-1793*. Carlton, Victoria: Melbourne University Press.

Hulot, Etienne-Gabriel-Joseph, baron. 1894. *D'Entrecasteaux, 1737-1793*. Paris: Société de géographie.

Jurien de la Gravière, Pierre-Roch. 1860. *Souvenirs d'un amiral*, 2 vols. Paris: Hachette.

Keeley, Lawrence H. 1996. *War Before Civilization*. New York & Oxford: Oxford University Press

Labillardière, Jacques-Julien Houtou de. 1800. *Relation du voyage à la recherche de la Pérouse*, 2 vols. Paris: H. J. Jansen

~ 1800. *Voyage in search of La Pérouse, performed by order of the Constituent Assembly, during the years 1791, 1792, 1793, and 1794*. London: J. Stockdale.

~ 1804-1806. *Novae Hollandiae plantarum specimen*, 2 vols. Paris: Dominae Huzard.

~ 1824-1825. *Sertum austro-caledonicum*, 2 parts. Paris: Dominae Huzard.

Lacaze, A. de. 1859. "La Billardière." *Nouvelle Biographie Générale* 28: 368-371. Paris: F. Didot frères.

La Pérouse. See Milet-Mureau 1797.

Lasègue, Antoine. 1845. *Musée botanique de Monsieur Benjamin Delessert.* Paris: Fortin, Masson.

Leschenault de La Tour, Jean-Baptiste-Louis-Claude-Théodore. 1807-1816. "Notice sur la Végétation de la Nouvelle-Hollande et de la Terre de Diemen." François Péron. *Voyage de découvertes aux Terres Australes . . . pendant les années 1800-1804*, 2: 358-372. Paris: Imprimerie Impériale.

Lesseps, Jean-Baptiste-Barthélemy, baron de. 1880. *Voyage de M. de Lesseps du Kamtchatka en France.* Paris: M. Dreyfous.

Lestringant, Frank. 1997. *Cannibals: The Discovery and Representation of the Cannibals from Columbus to Jules Verne.* Berkeley: University of California Press, 1997.

Letouzey, Yvonne, ed. 1989. *Le Jardin des Plantes à la croisée des chemins avec André Thouin, 1747-1824.* Paris: Muséum National d'Histoire Naturelle.

Liebersohn, Harry. 1996. "Recent Works on Travel Writing." *Journal of Modern History* 68 (September): 617-628.

Maiden, J. H. 1910. "Records of the Earlier French botanists as regards Australian Plants." *Journal of the Proceedings of the Royal Society of New South Wales* 44: 123-155.

Marcel, Gabriel-Alexandre. 1888. *Une Expédition oubliée à la recherche de Lapérouse.* Paris: L. Baudoin.

Milet-Mureau, Louis-Antoine Destouff, baron de. 1797. *Voyage de La Pérouse autour du monde, publié conformément au décret du 22 avril 1791*, 5 vols. Paris: Imprimerie de la République.

Miller, David Philip, and Peter Hanns Reill, eds. 1996. *Visions of Empire: Voyages, botany, and representations of nature.* Cambridge: Cambridge University Press.

Nelson, E. Charles. 1974. "The locations of collection and collectors of specimens described by Labillardière in *Novae Hollandiae Plantarum Specimen*: additional notes." *Papers & Proceedings of the Royal Society of Tasmania* 108: 159-170.

~ 1975 "The collectors and type locations of some of Labillardière's 'Terra van Leeuwin' (Western Australia) specimens." *Taxon* 24:319-336.

Niaussat, Pierre M. 1989. "Fleurieu et les naturalistes de son temps." Ulane Bonnel, ed. *Fleurieu et la marine de son temps.* Paris: Economica.

Plomley, Brian, and Josiane Piard-Bernier. 1993. *The General: the visits of the expedition led by Bruny d'Entrecasteaux to Tasmanian Waters in 1792 and 1793.* Launceston: Queen Victoria Museum.

Richard, Hélène. 1986. *Le Voyage d'Entrecasteaux à la recherche de Lapérouse.* Paris: Editions du Comité des Travaux Historiques et Scientifiques.

Rossel, Elisabeth-Paul-Edouard, chevalier de. 1808. *Voyage d'Entrecasteaux, envoyé à la recherche de La Pérouse.* 2 vols. Paris: Imprimerie impériale.

Saint-Hilaire, Auguste de. 1834. "Discours prononcé le 10 janvier 1834 sur la tombe de M. de La Billardière." *Annales des sciences naturelles. Botanique*. ser. 2. 1: 39-44.

Smith, Bernard William. 1985. *European Vision and the South Pacific, 1768-1850*, 2nd ed. New Haven & London: Yale University Press.

Stafleu, Frans A. 1967. *Adanson, Labillardière, De Candolle*. Lehre: J. Cramer.

Steenis, E. G. G. J. van, M. J. van Steenis-Kruseman, and E. A. Backer. 1954. "Louis-Auguste Deschamps, a prominent but ill-fated early explorer of the flora of Java, 1793-1798." *Bulletin of the British Museum (Natural History) Historical Series* 1, no. 2: 51-68.

Willis, James H. 1953. "Plants of the Recherche Archipelago, W.A." *Australian Geographical Society Reports* 1 (3a): 3-35.

General Index

Plant Index

ARCHIVES INTERNATIONALES D'HISTOIRE DES IDÉES
*
INTERNATIONAL ARCHIVES OF THE HISTORY OF IDEAS

1. E. Labrousse: *Pierre Bayle.* Tome I: *Du pays de foix à la cité d'Erasme.* 1963; 2nd printing 1984 ISBN 90-247-3136-4
   For Tome II *see below under Volume 6.*
2. P. Merlan: *Monopsychism, Mysticism, Metaconsciousness.* Problems of the Soul in the Neoaristotelian and Neoplatonic Tradition. 1963; 2nd printing 1969 ISBN 90-247-0178-3
3. H.G. van Leeuwen: *The Problem of Certainty in English Thought, 1630–1690.* With a Preface by R.H. Popkin. 1963; 2nd printing 1970 ISBN 90-247-0179-1
4. P.W. Janssen: *Les origines de la réforme des Carmes en France au 17e Siècle.* 1963; 2nd printing 1969 ISBN 90-247-0180-5
5. G. Sebba: *Bibliographia Cartesiana.* A Critical Guide to the Descartes Literature (1800–1960). 1964 ISBN 90-247-0181-3
6. E. Labrousse: *Pierre Bayle.* Tome II: *Heterodoxie et rigorisme.* 1964 ISBN 90-247-0182-1
7. K.W. Swart: *The Sense of Decadence in 19th-Century France.* 1964 ISBN 90-247-0183-X
8. W. Rex: *Essays on Pierre Bayle and Religious Controversy.* 1965 ISBN 90-247-0184-8
9. E. Heier: *L.H. Nicolay (1737–1820) and His Contemporaries.* Diderot, Rousseau, Voltaire, Gluck, Metastasio, Galiani, D'Escherny, Gessner, Bodmer, Lavater, Wieland, Frederick II, Falconet, W. Robertson, Paul I, Cagliostro, Gellert, Winckelmann, Poinsinet, Lloyd, Sanchez, Masson, and Others. 1965 ISBN 90-247-0185-6
10. H.M. Bracken: *The Early Reception of Berkeley's Immaterialism, 1710–1733.* [1958] Rev. ed. 1965 ISBN 90-247-0186-4
11. R.A. Watson: *The Downfall of Cartesianism, 1673–1712.* A Study of Epistemological Issues in Late 17th-Century Cartesianism. 1966 ISBN 90-247-0187-2
12. R. Descartes: *Regulæ ad Directionem Ingenii.* Texte critique établi par Giovanni Crapulli avec la version hollandaise du 17e siècle. 1966 ISBN 90-247-0188-0
13. J. Chapelain: *Soixante-dix-sept Lettres inédites à Nicolas Heinsius (1649–1658).* Publiées d'après le manuscrit de Leyde avec une introduction et des notes par B. Bray. 1966 ISBN 90-247-0189-9
14. C. B. Brush: *Montaigne and Bayle.* Variations on the Theme of Skepticism. 1966 ISBN 90-247-0190-2
15. B. Neveu: *Un historien à l'Ecole de Port-Royal.* Sébastien le Nain de Tillemont (1637–1698). 1966 ISBN 90-247-0191-0
16. A. Faivre: *Kirchberger et l'Illuminisme du 18e siècle.* 1966 ISBN 90-247-0192-9
17. J.A. Clarke: *Huguenot Warrior.* The Life and Times of Henri de Rohan (1579–1638). 1966 ISBN 90-247-0193-7
18. S. Kinser: *The Works of Jacques-Auguste de Thou.* 1966 ISBN 90-247-0194-5
19. E.F. Hirsch: *Damião de Gois.* The Life and Thought of a Portuguese Humanist (1502–1574). 1967 ISBN 90-247-0195-3
20. P.J.S. Whitemore: *The Order of Minims in 17th-Century France.* 1967 ISBN 90-247-0196-1
21. H. Hillenaar: *Fénelon et les Jésuites.* 1967 ISBN 90-247-0197-X
22. W.N. Hargreaves-Mawdsley: *The English Della Cruscans and Their Time, 1783–1828.* 1967 ISBN 90-247-0198-8
23. C.B. Schmitt: *Gianfrancesco Pico della Mirandola (1469–1533) and his Critique of Aristotle.* 1967 ISBN 90-247-0199-6
24. H.B. White: *Peace among the Willows.* The Political Philosophy of Francis Bacon. 1968 ISBN 90-247-0200-3

ARCHIVES INTERNATIONALES D'HISTOIRE DES IDÉES
*
INTERNATIONAL ARCHIVES OF THE HISTORY OF IDEAS

25. L. Apt: *Louis-Philippe de Ségur.* An Intellectual in a Revolutionary Age. 1969 ISBN 90-247-0201-1
26. E.H. Kadler: *Literary Figures in French Drama (1784–1834).* 1969 ISBN 90-247-0202-X
27. G. Postel: *Le Thrésor des prophéties de l'univers.* Manuscrit publié avec une introduction et des notes par F. Secret. 1969 ISBN 90-247-0203-8
28. E.G. Boscherini: *Lexicon Spinozanum.* 2 vols., 1970 Set ISBN 90-247-0205-4
29. C.A. Bolton: *Church Reform in 18th-Century Italy.* The Synod of Pistoia (1786). 1969 ISBN 90-247-0208-9
30. D. Janicaud: *Une généalogie du spiritualisme français.* Aux sources du bergsonisme: [Félix] Ravaisson [1813–1900] et la métaphysique. 1969 ISBN 90-247-0209-7
31. J.-E. d'Angers: *L'Humanisme chrétien au 17e siècle.* St. François de Sales et Yves de Paris. 1970 ISBN 90-247-0210-0
32. H.B. White: *Copp'd Hills towards Heaven.* Shakespeare and the Classical Polity. 1970 ISBN 90-247-0250-X
33. P.J. Olscamp: *The Moral Philosophy of George Berkeley.* 1970 ISBN 90-247-0303-4
34. C.G. Noreña: *Juan Luis Vives (1492–1540).* 1970 ISBN 90-247-5008-3
35. J. O'Higgens: *Anthony Collins (1676–1729), the Man and His World.* 1970 ISBN 90-247-5007-5
36. F.T. Brechka: *Gerard van Swieten and His World (1700–1772).* 1970 ISBN 90-247-5009-1
37. M.H. Waddicor: *Montesquieu and the Pilosophy of Natural Law.* 1970 ISBN 90-247-5039-3
38. O.R. Bloch: *La Philosophie de Gassendi (1592–1655).* Nominalisme, matérialisme et métaphysique. 1971 ISBN 90-247-5035-0
39. J. Hoyles: *The Waning of the Renaissance (1640–1740).* Studies in the Thought and Poetry of Henry More, John Norris and Isaac Watts. 1971 ISBN 90-247-5077-6
    For Henry More, *see also below under Volume 122 and 127.*
40. H. Bots: *Correspondance de Jacques Dupuy et de Nicolas Heinsius (1646–1656).* 1971 ISBN 90-247-5092-X
41. W.C. Lehmann: *Henry Home, Lord Kames, and the Scottish Enlightenment.* A Study in National Character and in the History of Ideas. 1971 ISBN 90-247-5018-0
42. C. Kramer: *Emmery de Lyere et Marnix de Sainte Aldegonde.* Un admirateur de Sébastien Franck et de Montaigne aux prises avec le champion des calvinistes néerlandais.[Avec le texte d'Emmery de Lyere:] *Antidote ou contrepoison contre les conseils sanguinaires et envinemez de Philippe de Marnix Sr. de Ste. Aldegonde.* 1971 ISBN 90-247-5136-5
43. P. Dibon: *Inventaire de la correspondance (1595–1650) d'André Rivet (1572–1651).* 1971 ISBN 90-247-5112-8
44. K.A. Kottman: *Law and Apocalypse.* The Moral Thought of Luis de Leon (1527?–1591). 1972 ISBN 90-247-1183-5
45. F.G. Nauen: *Revolution, Idealism and Human Freedom.* Schelling, Hölderlin and Hegel, and the Crisis of Early German Idealism. 1971 ISBN 90-247-5117-9
46. H. Jensen: *Motivation and the Moral Sense in Francis Hutcheson's* [1694–1746] *Ethical Theory.* 1971 ISBN 90-247-1187-8
47. A. Rosenberg: *[Simon] Tyssot de Patot and His Work (1655–1738).* 1972 ISBN 90-247-1199-1
48. C. Walton: *De la recherche du bien.* A study of [Nicolas de] Malebranche's [1638–1715] Science of Ethics. 1972 ISBN 90-247-1205-X

ARCHIVES INTERNATIONALES D'HISTOIRE DES IDÉES
*
INTERNATIONAL ARCHIVES OF THE HISTORY OF IDEAS

49. P.J.S. Whitmore (ed.): *A 17th-Century Exposure of Superstition.* Select Text of Claude Pithoys (1587–1676). 1972 ISBN 90-247-1298-X
50. A. Sauvy: *Livres saisis à Paris entre 1678 et 1701.* D'après une étude préliminaire de Motoko Ninomiya. 1972 ISBN 90-247-1347-1
51. W.R. Redmond: *Bibliography of the Philosophy in the Iberian Colonies of America.* 1972 ISBN 90-247-1190-8
52. C.B. Schmitt: *Cicero Scepticus.* A Study of the Influence of the *Academica* in the Renaissance. 1972 ISBN 90-247-1299-8
53. J. Hoyles: *The Edges of Augustanism.* The Aesthetics of Spirituality in Thomas Ken, John Byrom and William Law. 1972 ISBN 90-247-1317-X
54. J. Bruggeman and A.J. van de Ven (éds.): *Inventaire* des pièces d'Archives françaises se rapportant à l'Abbaye de Port-Royal des Champs et son cercle et à la Résistance contre la Bulle *Unigenitus* et à l'Appel. 1972 ISBN 90-247-5122-5
55. J.W. Montgomery: *Cross and Crucible.* Johann Valentin Andreae (1586–1654), Phoenix of the Theologians. Volume I: Andreae's Life, World-View, and Relations with Rosicrucianism and Alchemy; Volume II: The *Chymische Hochzeit* with Notes and Commentary. 1973 Set ISBN 90-247-5054-7
56. O. Lutaud: *Des révolutions d'Angleterre à la Révolution française.* Le tyrannicide & *Killing No Murder* (Cromwell, *Athalie*, Bonaparte). 1973 ISBN 90-247-1509-1
57. F. Duchesneau: *L'Empirisme de Locke.* 1973 ISBN 90-247-1349-8
58. R. Simon (éd.): *Henry de Boulainviller* – Œuvres Philosophiques, Tome I. 1973 ISBN 90-247-1332-3

    For Œuvres Philosophiques, Tome II *see below under Volume 70.*
59. E.E. Harris: *Salvation from Despair.* A Reappraisal of Spinoza's Philosophy. 1973 ISBN 90-247-5158-6
60. J.-F. Battail: *L'Avocat philosophe Géraud de Cordemoy (1626–1684).* 1973 ISBN 90-247-1542-3
61. T. Liu: *Discord in Zion.* The Puritan Divines and the Puritan Revolution (1640–1660). 1973 ISBN 90-247-5156-X
62. A. Strugnell: *Diderot's Politics.* A Study of the Evolution of Diderot's Political Thought after the *Encyclopédie.* 1973 ISBN 90-247-1540-7
63. G. Defaux: *Pantagruel et les Sophistes.* Contribution à l'histoire de l'humanisme chrétien au 16e siècle. 1973 ISBN 90-247-1566-0
64. G. Planty-Bonjour: *Hegel et la pensée philosophique en Russie (1830–1917).* 1974 ISBN 90-247-1576-8
65. R.J. Brook: [*George*] *Berkeley's Philosophy of Science.* 1973 ISBN 90-247-1555-5
66. T.E. Jessop: *A Bibliography of George Berkeley.* With: *Inventory of Berkeley's Manuscript Remains* by A.A. Lucc. 2nd revised and enlarged ed. 1973 ISBN 90-247-1577-6
67. E.I. Perry: *From Theology to History.* French Religious Controversy and the Revocation of the Edict of Nantes. 1973 ISBN 90-247-1578-4
68. P. Dibbon, H. Bots et E. Bots-Estourgie: *Inventaire de la correspondance (1631–1671) de Johannes Fredericus Gronovius* [1611–1671]. 1974 ISBN 90-247-1600-4
69. A.B. Collins: *The Secular is Sacred.* Platonism and Thomism in Marsilio Ficino's *Platonic Theology.* 1974 ISBN 90-247-1588-1

ARCHIVES INTERNATIONALES D'HISTOIRE DES IDÉES
*
INTERNATIONAL ARCHIVES OF THE HISTORY OF IDEAS

70. R. Simon (éd.): *Henry de Boulainviller.* Œuvres Philosophiques, Tome II. 1975 ISBN 90-247-1633-0
    For Œuvres Philosophiques, Tome I *see under Volume 58.*
71. J.A.G. Tans et H. Schmitz du Moulin: *Pasquier Quesnel devant la Congrégation de l'Index.* Correspondance avec Francesco Barberini et mémoires sur la misc à l'Index de son édition des Œuvres de Saint Léon, publiés avec introduction et annotations. 1974 ISBN 90-247-1661-6
72. J.W. Carven: *Napoleon and the Lazarists (1804–1809).* 1974 ISBN 90-247-1667-5
73. G. Symcox: *The Crisis of French Sea Power (1688–1697).* From the *Guerre d'Escadre* to the *Guerre de Course.* 1974 ISBN 90-247-1645-4
74. R. MacGillivray: *Restoration Historians and the English Civil War.* 1974 ISBN 90-247-1678-0
75. A. Soman (ed.): *The Massacre of St. Bartholomew.* Reappraisals and Documents. 1974 ISBN 90-247-1652-7
76. R.E. Wanner: *Claude Fleury (1640–1723) as an Educational Historiographer and Thinker.* With an Introduction by W.W. Brickman. 1975 ISBN 90-247-1684-5
77. R.T. Carroll: *The Common-Sense Philosophy of Religion of Bishop Edward Stillingfleet (1635–1699).* 1975 ISBN 90-247-1647-0
78. J. Macary: *Masque et lumières au 18e [siècle].* André-François Deslandes, Citoyen et philosophe (1689–1757). 1975 ISBN 90-247-1698-5
79. S.M. Mason: *Montesquieu's Idea of Justice.* 1975 ISBN 90-247-1670-5
80. D.J.H. van Elden: *Esprits fins et esprits géométriques dans les portraits de Saint-Simon.* Contributions à l'étude du vocabulaire et du style. 1975 ISBN 90-247-1726-4
81. I. Primer (ed.): *Mandeville Studies.* New Explorations in the Art and Thought of Dr Bernard Mandeville (1670–1733). 1975 ISBN 90-247-1686-1
82. C.G. Noreña: *Studies in Spanish Renaissance Thought.* 1975 ISBN 90-247-1727-2
83. G. Wilson: *A Medievalist in the 18th Century.* Le Grand d'Aussy and the Fabliaux ou Contes. 1975 ISBN 90-247-1782-5
84. J.-R. Armogathe: *Theologia Cartesiana.* L'explication physique de l'Eucharistie chez Descartes et Dom Robert Desgabets. 1977 ISBN 90-247-1869-4
85. Bérault Stuart, Seigneur d'Aubigny: *Traité sur l'art de la guerre.* Introduction et édition par Élie de Comminges. 1976 ISBN 90-247-1871-6
86. S.L. Kaplan: *Bread, Politics and Political Economy in the Reign of Louis XV.* 2 vols., 1976 Sct ISBN 90-247-1873-2
87. M. Lienhard (ed.): *The Origins and Characteristics of Anabaptism / Les débuts et les caractéristiques de l'Anabaptisme.* With an Extensive Bibliography / Avec une bibliographie détaillée. 1977 ISBN 90-247-1896-1
88. R. Descartes: *Règles utiles et claires pour la direction de l'esprit en la recherche de la vérité.* Traduction selon le lexique cartésien, et annotation conceptuelle par J.-L. Marion. Avec des notes mathématiques de P. Costabel. 1977 ISBN 90-247-1907-0
89. K. Hardesty: *The 'Supplément' to the 'Encyclopédie'.* [Diderot et d'Alembert]. 1977 ISBN 90-247-1965-8
90. H.B. White: *Antiquity Forgot.* Essays on Shakespeare, [Francis] Bacon, and Rembrandt. 1978 ISBN 90-247-1971-2
91. P.B.M. Blaas: *Continuity and Anachronism.* Parliamentary and Constitutional Development in Whig Historiography and in the Anti-Whig Reaction between 1890 and 1930. 1978 ISBN 90-247-2063-X

ARCHIVES INTERNATIONALES D'HISTOIRE DES IDÉES
*
INTERNATIONAL ARCHIVES OF THE HISTORY OF IDEAS

92. S.L. Kaplan (ed.): *La Bagarre.* Ferdinando Galiani's (1728–1787) 'Lost' Parody. With an Introduction by the Editor. 1979 ISBN 90-247-2125-3
93. E. McNiven Hine: *A Critical Study of [Étienne Bonnot de] Condillac's* [1714–1780] *'Traité des Systèmes'.* 1979 ISBN 90-247-2120-2
94. M.R.G. Spiller: *Concerning Natural Experimental Philosphy.* Meric Casaubon [1599–1671] and the Royal Society. 1980 ISBN 90-247-2414-7
95. F. Duchesneau: *La physiologie des Lumières.* Empirisme, modèles et théories. 1982 ISBN 90-247-2500-3
96. M. Heyd: *Between Orthodoxy and the Enlightenment.* Jean-Robert Chouet [1642–1731] and the Introduction of Cartesian Science in the Academy of Geneva. 1982 ISBN 90-247-2508-9
97. James O'Higgins: *Yves de Vallone* [1666/7–1705]: *The Making of an Esprit Fort.* 1982 ISBN 90-247-2520-8
98. M.L. Kuntz: *Guillaume Postel* [1510–1581]. Prophet of the Restitution of All Things. His Life and Thought. 1981 ISBN 90-247-2523-2
99. A. Rosenberg: *Nicolas Gueudeville and His Work (1652–172?).* 1982 ISBN 90-247-2533-X
100. S.L. Jaki: *Uneasy Genius: The Life and Work of Pierre Duhem* [1861-1916]. 1984 ISBN 90-247-2897-5; Pb (1987) 90-247-3532-7
101. Anne Conway [1631–1679]: *The Principles of the Most Ancient Modern Philosophy.* Edited and with an Introduction by P. Loptson. 1982 ISBN 90-247-2671-9
102. E.C. Patterson: *[Mrs.] Mary [Fairfax Greig] Sommerville* [1780–1872] *and the Cultivation of Science (1815–1840).* 1983 ISBN 90-247-2823-1
103. C.J. Berry: *Hume, Hegel and Human Nature.* 1982 ISBN 90-247-2682-4
104. C.J. Betts: *Early Deism in France.* From the so-called 'déistes' of Lyon (1564) to Voltaire's 'Lettres philosophiques' (1734). 1984 ISBN 90-247-2923-8
105. R. Gascoigne: *Religion, Rationality and Community.* Sacred and Secular in the Thought of Hegel and His Critics. 1985 ISBN 90-247-2992-0
106. S. Tweyman: *Scepticism and Belief in Hume's 'Dialogues Concerning Natural Religion'.* 1986 ISBN 90-247-3090-2
107. G. Cerny: *Theology, Politics and Letters at the Crossroads of European Civilization.* Jacques Basnage [1653–1723] and the Baylean Huguenot Refugees in the Dutch Republic. 1987 ISBN 90-247-3150-X
108. Spinoza's *Algebraic Calculation of the Rainbow & Calculation of Changes.* Edited and Translated from Dutch, with an Introduction, Explanatory Notes and an Appendix by M.J. Petry. 1985 ISBN 90-247-3149-6
109. R.G. McRae: *Philosophy and the Absolute.* The Modes of Hegel's Speculation. 1985 ISBN 90-247-3151-8
110. J.D. North and J.J. Roche (eds.): *The Light of Nature.* Essays in the History and Philosophy of Science presented to A.C. Crombie. 1985 ISBN 90-247-3165-8
111. C. Walton and P.J. Johnson (eds.): *[Thomas] Hobbes's 'Science of Natural Justice'.* 1987 ISBN 90-247-3226-3
112. B.W. Head: *Ideology and Social Science.* Destutt de Tracy and French Liberalism. 1985 ISBN 90-247-3228-X
113. A.Th. Peperzak: *Philosophy and Politics.* A Commentary on the Preface to Hegel's *Philosophy of Right.* 1987 ISBN Hb 90-247-3337-5; Pb ISBN 90-247-3338-3

ARCHIVES INTERNATIONALES D'HISTOIRE DES IDÉES
*
INTERNATIONAL ARCHIVES OF THE HISTORY OF IDEAS

114. S. Pines and Y. Yovel (eds.): *Maimonides* [1135-1204] *and Philosophy.* Papers Presented at the 6th Jerusalem Philosophical Encounter (May 1985). 1986 ISBN 90-247-3439-8
115. T.J. Saxby: *The Quest for the New Jerusalem, Jean de Labadie* [1610–1674] *and the Labadists (1610–1744).* 1987 ISBN 90-247-3485-1
116. C.E. Harline: *Pamphlets, Printing, and Political Culture in the Early Dutch Republic.* 1987 ISBN 90-247-3511-4
117. R.A. Watson and J.E. Force (eds.): *The Sceptical Mode in Modern Philosophy.* Essays in Honor of Richard H. Popkin. 1988 ISBN 90-247-3584-X
118. R.T. Bienvenu and M. Feingold (eds.): *In the Presence of the Past.* Essays in Honor of Frank Manuel. 1991 ISBN 0-7923-1008-X
119. J. van den Berg and E.G.E. van der Wall (eds.): *Jewish-Christian Relations in the 17th Century.* Studies and Documents. 1988 ISBN 90-247-3617-X
120. N. Waszek: *The Scottish Enlightenment and Hegel's Account of 'Civil Society'.* 1988 ISBN 90-247-3596-3
121. J. Walker (ed.): *Thought and Faith in the Philosophy of Hegel.* 1991 ISBN 0-7923-1234-1
122. Henry More [1614–1687]: *The Immortality of the Soul.* Edited with Introduction and Notes by A. Jacob. 1987 ISBN 90-247-3512-2
123. P.B. Scheurer and G. Debrock (eds.): *Newton's Scientific and Philosophical Legacy.* 1988 ISBN 90-247-3723-0
124. D.R. Kelley and R.H. Popkin (eds.): *The Shapes of Knowledge from the Renaissance to the Enlightenment.* 1991 ISBN 0-7923-1259-7
125. R.M. Golden (ed.): *The Huguenot Connection.* The Edict of Nantes, Its Revocation, and Early French Migration to South Carolina. 1988 ISBN 90-247-3645-5
126. S. Lindroth: *Les chemins du savoir en Suède.* De la fondation de l'Université d'Upsal à Jacob Berzelius. Études et Portraits. Traduit du suédois, présenté et annoté par J.-F. Battail. Avec une introduction sur Sten Lindroth par G. Eriksson. 1988 ISBN 90-247-3579-3
127. S. Hutton (ed.): *Henry More (1614–1687). Tercentenary Studies.* With a Biography and Bibliography by R. Crocker. 1989 ISBN 0-7923-0095-5
128. Y. Yovel (ed.): *Kant's Practical Philosophy Reconsidered.* Papers Presented at the 7th Jerusalem Philosophical Encounter (December 1986). 1989 ISBN 0-7923-0405-5
129. J.E. Force and R.H. Popkin: *Essays on the Context, Nature, and Influence of Isaac Newton's Theology.* 1990 ISBN 0-7923-0583-3
130. N. Capaldi and D.W. Livingston (eds.): *Liberty in Hume's 'History of England'.* 1990 ISBN 0-7923-0650-3
131. W. Brand: *Hume's Theory of Moral Judgment.* A Study in the Unity of *A Treatise of Human Nature.* 1992 ISBN 0-7923-1415-8
132. C.E. Harline (ed.): *The Rhyme and Reason of Politics in Early Modern Europe.* Collected Essays of Herbert H. Rowen. 1992 ISBN 0-7923-1527-8
133. N. Malebranche: *Treatise on Ethics* (1684). Translated and edited by C. Walton. 1993 ISBN 0-7923-1763-7
134. B.C. Southgate: *'Covetous of Truth'.* The Life and Work of Thomas White (1593–1676). 1993 ISBN 0-7923-1926-5
135. G. Santinello, C.W.T. Blackwell and Ph. Weller (eds.): *Models of the History of Philosophy.* Vol. 1: From its Origins in the Renaissance to the 'Historia Philosophica'. 1993 ISBN 0-7923-2200-2
136. M.J. Petry (ed.): *Hegel and Newtonianism.* 1993 ISBN 0-7923-2202-9

ARCHIVES INTERNATIONALES D'HISTOIRE DES IDÉES
*
INTERNATIONAL ARCHIVES OF THE HISTORY OF IDEAS

137. Otto von Guericke: *The New (so-called Magdeburg) Experiments* [Experimenta Nova, Amsterdam 1672]. Translated and edited by M.G. Foley Ames. 1994 ISBN 0-7923-2399-8
138. R.H. Popkin and G.M. Weiner (eds.): *Jewish Christians and Cristian Jews*. From the Renaissance to the Enlightenment. 1994 ISBN 0-7923-2452-8
139. J.E. Force and R.H. Popkin (eds.): *The Books of Nature and Scripture*. Recent Essays on Natural Philosophy, Theology, and Biblical Criticism in the Netherlands of Spinoza's Time and the British Isles of Newton's Time. 1994 ISBN 0-7923-2467-6
140. P. Rattansi and A. Clericuzio (eds.): *Alchemy and Chemistry in the 16th and 17th Centuries*. 1994 ISBN 0-7923-2573-7
141. S. Jayne: *Plato in Renaissance England*. 1995 ISBN 0-7923-3060-9
142. A.P. Coudert: *Leibniz and the Kabbalah*. 1995 ISBN 0-7923-3114-1
143. M.H. Hoffheimer: *Eduard Gans and the Hegelian Philosophy of Law*. 1995 ISBN 0-7923-3114-1
144. J.R.M. Neto: *The Christianization of Pyrrhonism*. Scepticism and Faith in Pascal, Kierkegaard, and Shestov. 1995 ISBN 0-7923-3381-0
145. R.H. Popkin (ed.): *Scepticism in the History of Philosophy*. A Pan-American Dialogue. 1996 ISBN 0-7923-3769-7
146. M. de Baar, M. Löwensteyn, M. Monteiro and A.A. Sneller (eds.): *Choosing the Better Part*. Anna Maria van Schurman (1607–1678). 1995 ISBN 0-7923-3799-9
147. M. Degenaar: *Molyneux's Problem*. Three Centuries of Discussion on the Perception of Forms. 1996 ISBN 0-7923-3934-7
148. S. Berti, F. Charles-Daubert and R.H. Popkin (eds.): *Heterodoxy, Spinozism, and Free Thought in Early-Eighteenth-Century Europe*. Studies on the *Traité des trois imposteurs*. 1996 ISBN 0-7923-4192-9
149. G.K. Browning (ed.): *Hegel's* Phenomenology of Spirit: *A Reappraisal*. 1997 ISBN 0-7923-4480-4
150. G.A.J. Rogers, J.M. Vienne and Y.C. Zarka (eds.): *The Cambridge Platonists in Philosophical Context*. Politics, Metaphysics and Religion. 1997 ISBN 0-7923-4530-4
151. R.L. Williams: *The Letters of Dominique Chaix, Botanist-Curé*. 1997 ISBN 0-7923-4615-7
152. R.H. Popkin, E. de Olaso and G. Tonelli (eds.): *Scepticism in the Enlightenment*. 1997 ISBN 0-7923-4643-2
153. L. de la Forge. Translated and edited by D.M. Clarke: *Treatise on the Human Mind (1664)*. 1997 ISBN 0-7923-4778-1
154. S.P. Foster: *Melancholy Duty*. The Hume-Gibbon Attack on Christianity. 1997 ISBN 0-7923-4785-4
155. J. van der Zande and R.H. Popkin (eds.): *The Skeptical Tradition Around 1800*. Skepticism in Philosophy, Science, and Society. 1997 ISBN 0-7923-4846-X
156. P. Ferretti: *A Russian Advocate of Peace. Vasilii Malinovskii (1765–1814)*. 1997 ISBN 0-7923-4846-6
157. M. Goldish: *Judaism in the Theology of Sir Isaac Newton*. 1998 ISBN 0-7923-4996-2
158. A.P. Coudert, R.H. Popkin and G.M. Weiner (eds.): *Leibniz, Mysticism and Religion*. 1998 ISBN 0-7923-5223-8
159. B. Fridén: *Rousseau's Economic Philosophy*. Beyond the Market of Innocents. 1998 ISBN 0-7923-5270-X
160. C.F. Fowler O.P.: *Descartes on the Human Soul*. Philosophy and the Demands of Christian Doctrine. 1999 ISBN 0-7923-5473-7

ARCHIVES INTERNATIONALES D'HISTOIRE DES IDÉES
*
INTERNATIONAL ARCHIVES OF THE HISTORY OF IDEAS

161. J.E. Force and R.H. Popkin (eds.): *Newton and Religion.* Context, Nature and Influence. 1999 ISBN 0-7923-5744-2
162. J.V. Andreae: *Christianapolis.* Introduced and translated by E.H. Thompson. 1999 ISBN 0-7923-5745-0
163. A.P. Coudert, S. Hutton, R.H. Popkin and G.M. Weiner (eds.): *Judaeo-Christian Intellectual Culture in the Seventeenth Century.* A Celebration of the Library of Narcissus Marsh (1638–1713). 1999 ISBN 0-7923-5789-2
164. T. Verbeek (ed.): *Johannes Clauberg* and Cartesian Philosophy in the Seventeenth Century. 1999 ISBN 0-7923-5831-7
165. A. Fix: *Fallen Angels.* Balthasar Bekker, Spirit Belief, and Confessionalism in the Seventeenth Century Dutch Republic. 1999 ISBN 0-7923-5876-7
166. S. Brown (ed.): *The Young Leibniz and his Philosophy (1646–76).* 2000 ISBN 0-7923-5997-6
167. R. Ward: *The Life of Henry More.* Parts 1 and 2. 2000 ISBN 0-7923-6097-4
168. Z. Janowski: *Cartesian Theodicy.* Descartes' Quest for Certitude. 2000 ISBN 0-7923-6127-X
169. J.D. Popkin and R.H. Popkin (eds.): *The Abbé Grégoire and his World.* 2000 ISBN 0-7923-6247-0
170. C.G. Caffentzis: *Exciting the Industry of Mankind. George Berkeley's Philosophy of Money.* 2000 ISBN 0-7923-6297-7
171. A. Clericuzio: *Elements, Principles and Corpuscles.* A Study of Atomisms and Chemistry in the Seventeenth Century. 2001 ISBN 0-7923-6782-0
172. H. Hotson: *Paradise Postponed.* Johann Heinrich Alsted and the Birth of Calvinist Millenarianism. 2001 ISBN 0-7923-6787-1
173. M. Goldish and R.H. Popkin (eds.): *Millenarianism and Messianism in Early Modern European Culture.* Volume I. Jewish Messianism in the Early Modern World. 2001 ISBN 0-7923-6850-9
174. K.A. Kottman (ed.): *Millenarianism and Messianism in Early Modern European Culture.* Volume II. Catholic Millenarianism: From Savonarola to the Abbé Grégoire. 2001 ISBN 0-7923-6849-5
175. J.E. Force and R.H. Popkin (eds.): *Millenarianism and Messianism in Early Modern European Culture.* Volume III. The Millenarian Turn: Millenarian Contexts of Science, Politics and Everyday Anglo-American Life in the Seventeenth and Eighteenth Centuries. 2001 ISBN 0-7923-6848-7
176. J.C. Laursen and R.H. Popkin (eds.): *Millenarianism and Messianism in Early Modern European Culture.* Volume IV. Continental Millenarians: Protestants, Catholics, Heretics. 2001 ISBN 0-7923-6847-9
177. C. von Linné: *Nemesis Divina.* (edited and translated with explanatory notes by M.J. Petry). 2001 ISBN 0-7923-6820-7
178. M.A. Badía Cabrera: *Hume's Reflection on Religion.* 2001 ISBN 0-7923-7024-4
179. R.L. Williams: *Botanophilia in Eighteenth-Century France.* The Spirit of the Enlightenment. 2001 ISBN 0-7923-6886-X
180. R. Crocker (ed.): *Religion, Reason and Nature in Early Modern Europe.* 2001 ISBN 1-4020-0047-2

ARCHIVES INTERNATIONALES D'HISTOIRE DES IDÉES
*
INTERNATIONAL ARCHIVES OF THE HISTORY OF IDEAS

181. E. Slowik: *Cartesian Spacetime*. Descartes' Physics and the Relational Theory of Space and Motion. 2001 ISBN 1-4020-0265-3
182. R.L. Williams: *French Botany in the Enlightenment*. The Ill-fated Voyages of *La Pérouse* and His Rescuers. 2003 ISBN 1-4020-1109-1
183. A. Leshem: *Newton on Mathematics and Spiritual Purity*. 2003 ISBN 1-4020-1151-2

KLUWER ACADEMIC PUBLISHERS – DORDRECHT / BOSTON / LONDON

FSC
www.fsc.org
MIX
Papier | Fördert
gute Waldnutzung
FSC® C083411

Kolibri 360 GmbH
Ferdinand-Jühlke-Straße 7
99095 Erfurt, Deutschland
produktsicherheit@kolibri360.de